T0190117

Environmental Footprints and Eco-design of Products and Processes

Series Editor

Subramanian Senthilkannan Muthu, Head of Sustainability - SgT Group and API, Hong Kong, Kowloon, Hong Kong

Indexed by Scopus

This series aims to broadly cover all the aspects related to environmental assessment of products, development of environmental and ecological indicators and eco-design of various products and processes. Below are the areas fall under the aims and scope of this series, but not limited to: Environmental Life Cycle Assessment; Social Life Cycle Assessment; Organizational and Product Carbon Footprints; Ecological, Energy and Water Footprints; Life cycle costing; Environmental and sustainable indicators; Environmental impact assessment methods and tools; Eco-design (sustainable design) aspects and tools; Biodegradation studies; Recycling; Solid waste management; Environmental and social audits; Green Purchasing and tools; Product environmental footprints; Environmental management standards and regulations; Eco-labels; Green Claims and green washing; Assessment of sustainability aspects.

Subramanian Senthilkannan Muthu
Editor

Environmental Assessment of Recycled Waste

 Springer

Editor
Subramanian Senthilkannan Muthu
Green Story Inc.
Toronto, ON, Canada

ISSN 2345-7651 ISSN 2345-766X (electronic)
Environmental Footprints and Eco-design of Products and Processes
ISBN 978-981-19-8325-2 ISBN 978-981-19-8323-8 (eBook)
https://doi.org/10.1007/978-981-19-8323-8

© The Editor(s) (if applicable) and The Author(s), under exclusive license to Springer Nature
Singapore Pte Ltd. 2023
This work is subject to copyright. All rights are solely and exclusively licensed by the Publisher, whether
the whole or part of the material is concerned, specifically the rights of translation, reprinting, reuse
of illustrations, recitation, broadcasting, reproduction on microfilms or in any other physical way, and
transmission or information storage and retrieval, electronic adaptation, computer software, or by similar
or dissimilar methodology now known or hereafter developed.
The use of general descriptive names, registered names, trademarks, service marks, etc. in this publication
does not imply, even in the absence of a specific statement, that such names are exempt from the relevant
protective laws and regulations and therefore free for general use.
The publisher, the authors, and the editors are safe to assume that the advice and information in this book
are believed to be true and accurate at the date of publication. Neither the publisher nor the authors or
the editors give a warranty, expressed or implied, with respect to the material contained herein or for any
errors or omissions that may have been made. The publisher remains neutral with regard to jurisdictional
claims in published maps and institutional affiliations.

This Springer imprint is published by the registered company Springer Nature Singapore Pte Ltd.
The registered company address is: 152 Beach Road, #21-01/04 Gateway East, Singapore 189721,
Singapore

This book is dedicated to:
The lotus feet of my beloved Lord
Pazhaniandavar
My beloved late Father
My beloved Mother
My beloved Wife Karpagam and Daughters-
Anu and Karthika
My beloved Brother
Last but not least
To everyone working towards the
Environmental Footprint Reduction and
Greening of Recycled Waste Sector

Preface

Waste Management is one of the main topics of discussion for an organization or even for a nation. Out of different destinations at the end of life for a product, recycling is absolutely the need of the hour and it is an inevitable destination. Literally all wastes (be it- post-industrial and post-consumer states) if cannot be reused, have to recycled and the significance of recycled products in the market has gained phenomenal importance.

Recycled products are in demand today and we are seeing many recycled alternatives for almost all industrial sectors. One of the million-dollar questions to answer in terms of recycling and recycled products is—whether the recycled products are environmentally sustainable than the virgin alternatives? It is highly imperative to ascertain the environmental footprints of recycled products and recycling processes and also find out the best possible ways to further improve the environmental benefits of such recycled products and recycling processes. This book, revolves around the environmental assessment of recycled waste with six interesting chapters.

Chapter "Carbon Footprint of Pipe Production Using Waste Plastics" deals around an interesting topic, i.e. Carbon Footprint of Pipe Production using Waste Plastics. This chapter discusses a Carbon Footprint case study to quantitatively analyze the three waste plastic pipes P1 (2 inches inner diameter, 4 mm wall thickness), P2 (3 inches inner diameter, 5 mm wall thickness), and P3 (4 inches inner diameter, 6 mm wall thickness) production process. From the analysis carried out, it is suggested that the use of waste plastic in pipe production is environment friendly and economically beneficial. Further, it also helps to enhance local job opportunities and improve social sustainability.

Chapter "Enviro-Economic Assessment of Plastic Cell-Filled Concrete Pavement" revolves around an important topic, which is, Enviro-economic Assessment of Plastic Cell-filled Concrete Pavement. This interesting chapter presents the methodology to assess the Ecological Footprint and as well as economic assessment of a plastic (waste) cell-filled concrete pavement construction. Deeper analysis from this study indicated that the plastic cell-filled concrete pavement is eco-friendly and cost-effective compared with the conventional pavements (for low traffic volumes).

Chapter "Carbon Footprint and Economic Assessment of LED Bulbs Recycling" is dedicated to deal with the Carbon Footprint and Economic Assessment of LED bulbs Recycling. This chapter presents the case study of environmental and economic assessments of recycling of LED bulbs. Very interesting results and meaningful conclusions are drawn from this study.

Chapter "Ecological Footprint Assessment of e-Waste Recycling" deals with an Empirical Investigation of Waste Management and Ecological Footprints in OECD Countries. This chapter presents a contribution to the literature by investigating the long-term effects of municipal waste, recycling, and landfills on the ecological footprint in 17 OECD countries from 1995–2018. This study investigates the impact of waste management on environmental degradation by using municipal waste per capita, the rate of municipal waste sent to landfills, recycling rate, per capita income, renewable energy consumption, trade openness, and ecological footprint per capita from 1995 to 2018 in OECD countries. Fully modified ordinary least squares (FMOLS) and Pooled Mean Group (PMG) estimation tests are employed to test this relationship. According to the findings arising from the study, the long-term coefficient shows that the amount of municipal waste per capita increases the ecological footprint per capita. On the other hand, the amount of municipal waste sent to landfills, recycling rate, and renewable energy consumption reduce the ecological footprint per capita.

Chapter "An Empirical Investigation of Waste Management and Ecological Footprints in OECD Countries" discusses around the Ecological Footprint Assessment of e-Waste Recycling. This study proposes a methodology to assess the Ecological Footprint (EF) of e-Waste recycling. As per the results derived from the study, it's found that the carbon absorption land contributes the highest among all other bio-productive lands for the e-Waste recycling.

Chapter "Ecological Footprint Assessment of Concrete Using e-Waste" deals with the Ecological Footprint Assessment of Concrete using e-Waste. In this study, the environmental assessment of concrete using e-Waste is examined and also the Sustainable Recycling Index (SRI) is also developed for the environmental assessment of e-Waste as a replacement for aggregate in plain cement concrete. Results from the study aid to have a conclusion—the production of concrete using e-Waste provides a sustainable option for e-Waste assimilation.

I take this opportunity to thank all the contributors for their earnest efforts to bring out this book successfully. I am sure readers of this book will find it very useful.

With best wishes,
Toronto, Canada   Dr. Subramanian Senthilkannan Muthu
December 2022

Contents

About the Editor

Dr. Subramanian Senthilkannan Muthu currently works at Green Story as Chief Sustainability Officer, Canada, and is based out of Hong Kong. He earned his PhD from The Hong Kong Polytechnic University, and is a renowned expert in the areas of Environmental Sustainability in Textiles & Clothing Supply Chain, Product Life Cycle Assessment (LCA) and Product Carbon Footprint Assessment (PCF) in various industrial sectors. He has five years of industrial experience in textile manufacturing, research and development and textile testing and over a decade's of experience in life cycle assessment (LCA), carbon and ecological footprints assessment of various consumer products. He has published more than 100 research publications, written numerous book chapters and authored/edited over 100 books in the areas of Carbon Footprint, Recycling, Environmental Assessment and Environmental Sustainability.

Carbon Footprint of Pipe Production Using Waste Plastics

Sajid Naeem, Dilawar Husain, Kirti Tewari, Nayab Zafar, Md Tanwir Alam, and Naveed Hussain

Abstract Recycling plastics have a significant effect on the reduction of greenhouse gases and enable to achieve sustainable goals prescribed by the United Nations. Finding applications to replace virgin plastic with recycled plastic without affecting functionality is a key problem. The pipe manufacturing plant comes under the small- and medium-scale industries. Plastic pipes are commonly used in irrigation, water supply, and building drainage, etc., They help to assimilate waste plastic and to procure the eco-system of the planet. In this study, the Carbon Footprint analysis is applied to quantitatively analyze the three waste plastic pipes P_1 (2 inches inner diameter, 4 mm wall thickness), P_2 (3 inches inner diameter, 5 mm wall thickness), and P_3 (4 inches inner diameter, 6 mm wall thickness) production process. The total emission for production of P_1 pipe is 0.292–0.451 $kgCO_2$ per meter of pipe; P_2 pipe is 0.360–0.598 $kgCO_2$ per meter of pipe and P_3 pipe is 0.449–0799 $kgCO_2$ per meter of pipe. The study suggested the use of recycled plastic in pipe production is environment friendly and economically beneficial. It also helps to enhance local job opportunities and improve social sustainability.

Keywords Carbon footprint · Waste plastics · Sustainable development · Pipe production · Recycling · Municipal solid waste

S. Naeem
Department of Applied Sciences, Maulana Mukhtar Ahmad Nadvi Technical Campus Malegaon, Malegaon 423203, India

D. Husain · M. T. Alam · N. Hussain
Department of Mechanical Engineering, Maulana Mukhtar Ahmad Nadvi Technical Campus Malegaon, Malegaon 423203, India

K. Tewari (✉)
Department of Mechanical Engineering, National Institute of Technology Sikkim, Ravangla 737139, India
e-mail: kirti@nitsikkim.ac.in

N. Zafar
Department of Mechanical Engineering, Adsul's Technical Campus, Ahmednagar 414005, India

© The Author(s), under exclusive license to Springer Nature Singapore Pte Ltd. 2023
S. S. Muthu (ed.), *Environmental Assessment of Recycled Waste*,
Environmental Footprints and Eco-design of Products and Processes,
https://doi.org/10.1007/978-981-19-8323-8_1

1

1 Introduction

Depending on the utility sought, plastics are easily sculpted into different shapes. Additionally, the mixture of monomers that are joined to other identical subunits to create a polymer is what constitutes plastics [1]. In modern society plastics are becoming an integral part of life and causing unrestrained accruement of plastic waste. Thus, its disposal has become the biggest challenge these days, as it contaminates the ecological system [2]. It creates a negative impact on the earth, living things, healthcare and wildlife. The plastic wastes are either damped in landfilled or burned with other wastes which are in both ways hazardous to the environment and contribute toward the global warming. Approximately 400 million tons of plastic wastes are produced per year and it is expected that the plastic waste insert into the aquatic ecosystem would be nearly 29 million by 2040 [3].

To reduce this uncontrolled and damaging growth of wastes, recycling is considered the eco-friendliest solution [4, 5]. Recycling reduces: (i) the further addition of plastic waste into the environment, (ii) the need for landfill, and (iii) energy consumption. Additionally, recycled plastics may be seen as a sustainable supply of raw materials and they can lessen carbon footprints and greenhouse gas (GHG) emissions. Recycling benefits the environment in addition to saving 20 to 50 percent on the cost of raw materials [6].

The production of virgin plastic is solely responsible for the pollution of the consumers [7]. Thus, to control the furtherance of plastic waste, there should also be a check on plastic manufacturing industries. To sustain environment, the plastic manufacturing industries should also promote recycling. The waste plastics are contaminating the ecosystem by 14% incinerate and 10% recycle as reported by the UN Environment Program 2022 Report. The UN Environment Program reported that 9.2 billion tonnes of plastic have been produced in the year 1950 to 2017. If plastic production is continued at the same rate, it may reach up to 1100 million tonnes per year estimated over the globe [8].

Plastics are available in different forms such as synthetic, semi-synthetics, and polymers etc. There are two major processes of recycling plastic waste: mechanical process and chemical process [9]. In mechanical recycling, the process starts with the melting of plastic, solidification into a plastic solid rock, cutting into small pieces for making granulates, and final production of new products. While in chemical recycling, plastic waste converts polymers into hydrocarbon molecules which is used as fuel and industrial chemicals. The present study considers only the mechanical recycling process of waste plastics and plastic pipes as a final product.

1.1 Plastics Pipe Manufacturing

The plastics pipes production (hollow cylindrical shaped tube) is as shown in Fig. 1. There are different types of plastics pipes available in the market such as

Waste Plastic Pipes

Fig. 1 Image of waste plastic pipes (self-made)

solid wall pipe, structured wall pipe, and barrier pipe. The manufacturing process includes extrusion, heating, melting, measurement, cooling, printing, and cutting at the assembly line. Pipe manufacturing can be generally classified into the following four steps:

1. Plastics Wastage to Plastics Solid Bricks
2. Plastics Solid Bricks to Small Pieces of Plastics Blocks
3. Plastics Blocks to Plastics Sands
4. Plastics Sands to Plastics Pipe.

Production of these pipes can be majorly derived from plastic wastes. The essential steps involved are heating, melting, mixing, and cooling to transform raw plastic material into a pipe shape. Depending on the application, plastic pipes can be manufactured in different diameters. An assembly line machine is used for plastic solid wall pipe production from waste plastic (raw material) as shown in Fig. 2. As discussed above the processes involved in manufacturing a pipe from the plastic waste, each stage can be analyzed for their energy demand and carbon footprint.

The study of a product or the system's energy requirements from raw material extraction to manufacturing, usage, and disposal is known as its "carbon footprint," and it is one of the crucial factors in determining how successful a process is in terms of its impact on the environment [6]. Thus, an attempt has been made in the present study to analyze the environmental impact of the process of recycling plastic waste into plastic pipes. The acquired knowledge can be comprehensively implemented by the plastic waste recycling industry for the further reduction of carbon footprints and analyzing of the emission value generated at each stage of the production process.

Extrusion Machine

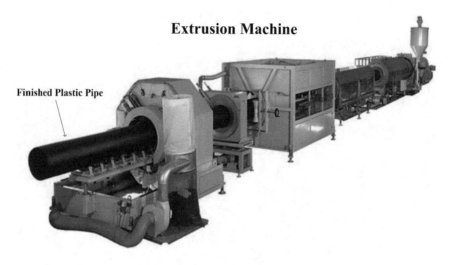

Fig. 2 Waste plastic to plastic pipe manufacturing machine (self-made)

2 Methodology

2.1 Case Study Selection

The study investigations are made in the city of Malegaon (20.5579° N, 74.5089° E), India. Malegaon city is located at the convergence of the Mosam and the Girna rivers. It is chosen as a case study because manufacturing pipes through waste plastic is the second largest industry in the city after the textile industry. Therefore, Malegaon city is a suitable location to fulfil the purposes of this research work. A database is created to record the production of waste plastic pipes. The raw material required to produce pipe is waste plastic collected from the waste dumping ground of the city.

According to the survey data for 2021, the annual municipal solid waste amount collected by the Malegaon municipal corporation was 61,320 tons. According to the solid waste characterization estimate made for Malegaon city (see Table 1), waste plastics contribute 13.25% of the total municipal solid waste. It indicates that waste plastic recycling significantly reduces the environmental impact of municipal solid waste in the city. The density of recycled plastic is 910–970 kg/m^3 which is used in manufacturing plastic products in Malegaon city. Generally, three types of waste plastic pipes are manufactured in Malegaon city, the details are mentioned in Table 2.

Table 1 Characterization of municipal solid waste (Malegaon city)

Solid waste	Contribution (%)
Organic material	59.25
Paper	7.25
Plastic	13.25
Metal	1.25
Glass	2
Leather and Rubber	2.75
Textile	6.25
Inert miscellaneous	9.25

Table 2 Details of pipes manufactured in Malegaon

Pipe types	Inner diameter (inches)	Wall thickness (mm)	Density (kg/m^3)	Weight (kg/m)
P$_1$	2	4	910–970	0.297–0.317
P$_2$	3	5	910–970	0.560–0.598
P$_3$	4	6	910–970	0.90–0.976

2.2 Production of Waste Plastic Pipes

The plastics wastes are collected from various sources in the form of plastic food wrappers, plastic bags, plastic beverage bottles, plastic grocery bags, straws, stirrers, plastic packaging, etc. There are a series of stages of recycling plastics waste. The steps involved are pellet production through waste plastic as shown in Fig. 3. Collecting, sorting, and refining the plastic so that it is ready to be utilized in new goods are some of these stages.

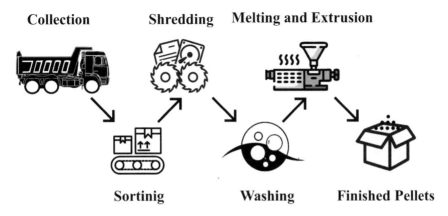

Fig. 3 Different stages of waste plastic pipe manufacturing (self-made)

- **Collection:** The waste plastics are collected from the end-users in a recycling container.
- **Sorting:** Different types of plastics are separated from the container
- **Reprocessing:** The plastic after processing is cleaned, powdered into flakes, heated, and then extruded to create fresh pellets.

Details of recycling steps

1. **Collection**

   It is the first stage of the recycling process of waste plastics. It involves the collection of recycling from public places, outside homes and schools, etc. The waste plastics are collected by a local administration either directly or using a waste management contractor.

   The other method of collection of plastic waste includes recycling centers and front of the store or local recycling sites. The collection of plastic is key for the recycling system to operate well. The more the suitable plastic for recycling is collected, the more the material available to be reprocessed and used back into new products.

2. **Sorting**

   In the second step, plastic is separated from other materials in a material recovery facility (MRF). Once this material has been further sorted into the various forms of plastic, it may be sent to a Plastic Recovery Facility (PRF). The mixed recyclables will be mechanically loaded onto a conveyer belt after being originally taken from the collecting vehicles. The steady flow of garbage moving through the sorting plant is maintained by conveyor belts. The material is separated into pieces ready for further processing using several procedures. Municipal solid waste management of Malegaon city used a sorting machine (12 hp capacity and rate of 15 tons municipal solid waste per hour) to sort waste plastic from waste. The techniques used in practice will vary by facility.

3. **Washing**

   The recycled plastic is cleared of impurities. Adhesives, leftover garbage from containers, food waste, and labels may all be removed by washing. These must be taken out, and cleaning the material also increases the quality of the recycled garbage. At this point, domestic activities can have a significant impact. The entire recycling system can function more efficiently by just immediately washing the plastic to remove some of the food or other things before they get dry and attach more strongly.

4. **Shredding/Grinding**

   Products made of plastic are broken down into smaller bits. Plastic must be crushed or shredded into tiny flakes and this is a crucial step in the recycling process. The cleaned and sorted plastic is fed through shredders where it is broken up into smaller bits. Depending on the shredder's categorization and techniques, the plastic is shredded in various ways. Hammer Mills are an illustration of a technique; they used to pulverize polymers in a rotational drum employing swiveling hammers. Other instances are shear shredders, which cut polymers to specifications required by the industry using guillotines and rotary cutters. To guarantee the creation of a pure stream of material, additional sorting may be necessary.

5. **Extrusion**

   Extrusion is the procedure by which plastic is melted and forced through an extruder. As the plastic exits the extruder, it is chopped into pellets. These crushed small pieces are converted into plastic rocks and further cut into pieces of plastic bricks. The plastic bricks are pushed into a crusher machine to cut into tiny pieces of plastics. These tiny solid plastics/sand size plastics are again processed by an extruder machine. New pellets are created by melting and extruding plastic.

   The extruder operates such as mixing, heating, cooling, and then shaping into pipes by dies. The plastic pipes are manufactured from the waste plastics and are recycled. Plastic manufacturing machine installation requires 150 square feet of space. The two operators are required to operate the machine. Plastics pipe production is the one of the best trades for recycling plastics and efforts to reduce plastic pollution.

2.3 Carbon Footprint Calculation

A life cycle perspective (such as embodied energy or embodied emissions) has been assessed to calculate the Carbon Footprint of product/human activities. For calculating the Carbon Footprint of waste plastic pipe manufacturing, a set of equations has been used in this study. The direct emissions due to fossil fuels used in machines for pipe manufacturing and transport vehicles are estimated as Eq. (1). The indirect emissions due to electricity used in machines for pipe manufacturing are estimated as Eq. (2). According to the IPCC (2006) program report [10], the suggested formula for assessment of GHG is demonstrated in Eqs. (3) and (4).

$$\text{Direct emission (fossil fuel use)} = \text{fuel use x emission factor of fuel} \quad (1)$$

$$\text{Indirect emission (electricity use)} = \text{electricity use x emission factor of electricity} \quad (2)$$

$$C = \sum_{i,j,k} AD_{i,j,k} EF_{i,j,k} \quad (3)$$

$$EF_{i,j,k} = c_k \eta_{i.j.k} \frac{44}{12} \quad (4)$$

where C is the amount of CO_2 emissions, $AD_{i,j,k}$ is the level of activity, $EF_{i,j,k}$ is the emission factor, i is the industry and region, j is the equipment and technology used, k is the type of fuel used, C_k is the carbon content, and $\eta_{i,j,k}$ is the oxidation rate.

For Carbon Footprint calculations, Eq. (5) has been used to calculate waste plastic collection, transportation of waste plastic from collection site to pipe industry, and pipe manufacturing processes.

$$CF_P = CF_c + CF_t + CF_m \tag{5}$$

where CF_c is the embodied carbon emissions of waste plastic, CF_t is the carbon emissions at the transportation of waste plastic form collection to pipe industry, CF_m is the carbon emissions at waste pipe manufacture.

3 Results and Discussion

3.1 *Plastic Pipe Production Process*

For the waste plastic pipes, waste plastics are collected from the waste dumping ground of the city and then transported to the different pipe manufacturing industries. Carbon Footprint per kg of virgin Polyvinyl chloride (PVC), Polypropylene (PP) and Polyethylene (PE) are 67 $kgCO_2$, 61.5 $kgCO_2$ and 58.7 $kgCO_2$, respectively [11–13]. The total emission for production of P_1 pipe is 0.292–0.451 $kgCO_2$ per meter; P_2 pipe is 0.360–0.598 $kgCO_2$ per meter and P_3 pipe is 0.449–0799 $kgCO_2$ per meter.

3.1.1 Collection

Sorting of MSW into waste plastics and organic/biodegradable components at the waste facility center is an energy intensive process. The energy use for on-site equipment is responsible for the GHG emissions. Sorting and segregation machine (capacity 12 hp; 15-ton municipal solid waste per hour) consumed 9.85 kWh electrical energy shown in Fig. 4. The percentage of waste plastic contribution is about 13.35% of the total municipal solid waste of the city (see Fig. 4). The estimated emission of waste plastic collection is about 0.107 $kgCO_2$ per kg of waste plastic (emission factor of electricity in India 0.82 tCO_2/MWh [14]).

3.1.2 Transportation

The waste dumping ground is located at the outskirts of the Malegaon city. Therefore, waste plastic transportation needed 4–10 km distance to reach pipe industry. Different types of transportation vehicles are mentioned in Table 3. Different types of vehicles are used in Malegaon city to transport waste plastics form collection site to pipe industries. The emissions due to the transportation of waste plastic are in the range of 2.97–20.8 $kgCO_2$/tonne-km.

Fig. 4 Sorting and collection machine of municipal solid waste in Malegaon dumping ground (self-made)

Table 3 The details of the transportation system and waste segregation machines

Vehicle	Capacity (in Tonne)	Fuel consumption rate (km/l)	Emission factor	Emission kgCO$_2$/tonne-km
Tractor-trolly	1–1.5	8	3.17 kgCO$_2$ per kg of diesel	13.86–20.8
Three- wheeler	1	35	3.17 kgCO$_2$ per kg of diesel	90.9
Tippers	7	8	3.17 kgCO$_2$ per kg of diesel	2.97
Heavy duty truck with hydraulic containers	3	10	3.17 kgCO$_2$ per kg of diesel	8.67
Solid waste segregation machine	15 Tonne/hr	12 Electric Motors, capacity 1hp each motor	0.82 tCO$_2$ per MWh	4.9×10^{-4} tCO$_2$/tonne-hr

3.1.3 Manufacturing (Extrusion)

Waste plastic pellets are used in the hooper of the second extruder machine. The collection of plastics wastes is pushed into the hooper of the extruder machine as input (see Fig. 5). The input plastic waste of 100 kg is processed per hour by the extruder machine screw rotated by AC motors (20 kW capacity; emission factor of electricity generation in India is about 0.82 tCO_2/MWh [14]) and heating and cooling processes required to manufacturing pipes. The waste plastics are melted by the electric heater at above 200 °C. The melted plastics flow through the water-cooling system to form solid plastic rocks (see Fig. 6a). Plastic rocks are further converted into tiny pellets (see Fig. 6b).

The overall electricity consumed by extrusion machine is 50–60 kWh/hr. The production rates of P1, P2, and P3 pipes are 140–160 m; 120–140 m and 100- 120 m; respectively. The emission during the operation of the extrusion machine to produce P_1 pipe is 0.256–0.351 $kgCO_2$ per meter of pipe; P_2 pipe is 0.293–0.410 $kgCO_2$ per meter of pipe and P_3 pipe is 0.242–0.492 $kgCO_2$ per meter of pipe. The collective emissions in all the three processes of pipe P1, P2, and P3 have been shown in Fig. 7. It indicates that the extrusion process has contributed maximum environmental impact

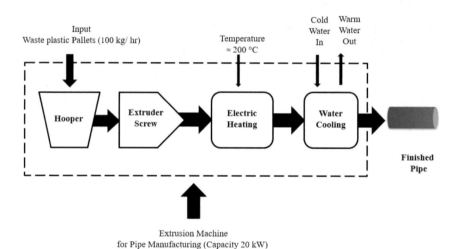

Fig. 5 Extrusion process involves to the manufacturing of pipe (self-made)

Fig. 6 a Image of waste plastic rock; **b** image of tiny pellets (self-made)

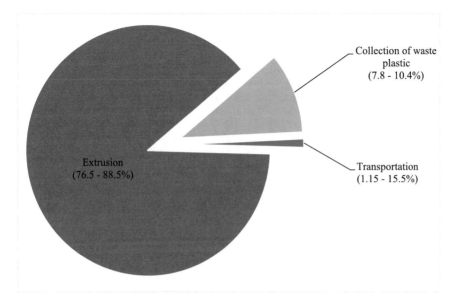

Fig. 7 Carbon Footprint distribution of pipe production (self-made)

(i.e., 76.5–88.5% of the total Carbon Footprint of pipe production) followed by the shorting process (i.e., 7.8–10.4%) of waste plastic production.

4 Conclusions

The Carbon Footprint of waste plastic pipes manufacturing has been examined for the production unit situated in Malegaon, Maharashtra, India. Based on the activities involved at the various stages of pipe production, the product-related GHG emissions have been classified under three stages, namely collection of waste plastics stage, transportation of waste plastics from the dumping ground to production unit/manufacturing plants, and extrusion process to make finished products (pipe). Results revealed that total emissions produced by the manufacturing of P_1, P_2 and P_3 pipes are (0.292–0.451) $kgCO_2$/meter, (0.360–0.598) $kgCO_2$/meter and (0.449–0799) $kgCO_2$/meter, respectively. The breakdown of the Carbon Footprint shown under various stages of pipe production reveals that the largest source of emission is the extrusion process followed by waste plastics collection. The remaining emission results from the transportation of waste plastics.

The results obtained by using the Carbon Footprint indicator provide an environmental assessment of pipe production and it helps to develop an assessment of sustainable recycling limits for waste products. The manufacturing industry may reduce its Carbon Footprint by using renewable energy sources, it also helps to improve industrial sustainability.

References

1. Evode N, Qamar SA, Bilal M, Barceló D, Iqbal HMN (2021) Plastic waste and its management strategies for environmental sustainability. Case Stud Chem Environ Eng 4:100142
2. Abid AYZ, Usama AA, Husain D, Sharma M, Prakash R (2022). Ecological footprint assessment of recycled asphalt pavement construction. In: Muthu SS (ed) Environmental footprints of recycled products. Environmental footprints and eco-design of products and processes. Springer, Singapore. https://doi.org/10.1007/978-981-16-8426-5_5
3. Environment, U. N., drowning in plastics – marine litter and plastic waste vital graphics. UNEP - UN Environment Programme 2021. https://www.unep.org/resources/report/drowning-plastics-marine-litter-and-plastic-waste-vital-graphics
4. Husain D, Tewari K, Sharma M, Ahmad A, Prakash R (2022) Ecological footprint of multi-silicon photovoltaic module recycling. In: Muthu SS (ed) Environmental footprints of recycled products. Environmental footprints and eco-design of products and processes. Springer, Singapore
5. Nawandar V, Husain D, Prakash R (2021) Ecological footprint assessment and its reduction for packaging industry. Assessment of ecological footprints, pp 41–78. https://doi.org/10.1080/19397038.2020.1783719
6. Finnveden G (1999) Methodological aspects of life cycle assessment of integrated solid waste management systems. Resour Conserv Recycl 26(3–4):173–187
7. Thompson RC, Moore CJ, Saal FS, Swan SH (2009) Plastics, the environment, and human health: current consensus and future trends. Philos Trans R Soc Lond B Biol Sci 364(1526):2153–2166. https://doi.org/10.1098/rstb.2009.0053.PMID:19528062;PMCID:PMC2873021
8. United Nations Environment Programme (UNEP) "Plastic Pollution" 2022. https://www.unep.org/plastic-pollution
9. Ragaert K, Delva L, Van Geem K (2017) Mechanical and chemical recycling of solid plastic waste. Waste Manag 69:24–58
10. Intergovernmental Panel on Climate Change (IPCC) (2006) IPCC guidelines for national greenhouse gas inventories, vol 4 agriculture forestry and other land use.
11. Camaratta R, Volkmer TM, Osorio AG (2020) Embodied energy in beverage packaging J Environ Manage
12. Al-Nuaimi S, Banawi AAA, Al-Ghamdi SG (2009) Environmental and economic life cycle analysis of primary construction materials sourcing under geopolitical uncertainties: a case study of Qatar Sustain 11:6000
13. Narita N, Sagisaka M, Inaba A (2002) Life cycle inventory analysis of CO_2 emissions: manufacturing commodity plastics in Japan. Int J Life Cycle Assess
14. Ministry of Power Central Electricity Authority, Government of India (MPCEA) (2016) "CO_2 baseline database for the Indian power sector, user guide 2016". Accessed November 2021

Enviro-Economic Assessment of Plastic Cell-Filled Concrete Pavement

Ansari Tauseef, Naveed Akhtar, Dilawar Husain, Roshan Ambadas Birari, Shilpa Ravindra Khedkar, Sachin Bhaskar Khairnar, Abdullah Khan, A. S. A. Ansari Ismail, and Mohammed Junaid

Abstract The plastic cell-filled concrete pavement (rigid pavement) consists of a framework of plastic cells over the compacted subgrade/sub-base, filled with concrete which has proved to be a very promising solution for overloaded vehicles, inadequate drainage facilities, and waterlogging problems. The technology also helps to achieve the sustainability goals in road construction. The study develops a methodology to assess the Ecological Footprint of a plastic (waste) cell-filled concrete pavement construction. The Ecological Footprint of plastic cell-filled pavement construction (3.75 m width, 100 mm thickness and 1 km length) is estimated as 22–23.45 gha. The Ecological Footprint of per square meter surface area of the plastic cell-filled concrete pavement construction is about 0.0058–0.0062 gha. The economic assessment of the plastic cell-filled concrete pavement construction has been estimated as estimated as Rs 1.47–1.53 million. The enviro-economic assessment of the plastic cell-filled concrete pavement indicated that the plastic cell-filled concrete pavement is eco-friendly and cost-effective compared with the conventional pavements (for low traffic volumes).

Keywords Ecological footprint · Rigid pavement · Sustainability · Recycling · Waste plastic management

1 Introduction

The global development in the construction activities requires a boost nowadays. Mass concrete is required to fulfil the needs of the transportation industry. The

A. Tauseef · N. Akhtar · R. A. Birari · S. R. Khedkar · S. B. Khairnar · A. Khan ·
A. S. A. Ismail · M. Junaid
Department of Civil Engineering, Maulana Mukhtar Ahmad Nadvi Technical Campus, Malegaon, Maharashtra 423203, India

D. Husain (✉)
Department of Mechanical Engineering, Maulana Mukhtar Ahmad Nadvi Technical Campus, Malegaon, Maharashtra 423203, India
e-mail: dilawar4friend@gmail.com

© The Author(s), under exclusive license to Springer Nature Singapore Pte Ltd. 2023
S. S. Muthu (ed.), *Environmental Assessment of Recycled Waste*,
Environmental Footprints and Eco-design of Products and Processes,
https://doi.org/10.1007/978-981-19-8323-8_2

concrete mass requires large number of natural aggregates (around 70–80%) and cement as a binder. The bituminous hot mix roads construction used previously have problems in rainy seasons due to weak drainage facility and less durability that will reduce life. Another option is the roads from high performance concrete having more durability but the initial cost of construction is very high [1]. The major challenges for the concrete roads are its flexural strength, durability and abrasion resistance properties. The aggregate from the waste plastic has all the above discussed properties [2]. Also, plastic use creates many environmental issues related to dumping, recycling and composting [3]. but due to compatibility issues in bonding of plastic with cement mixture, it leads to reduced strength and introduces more voids [4]. Civil agencies and Governments Authorities imposed ban on use of plastics which are not recyclable or reusable. But still the manufacturing continues in many parts of the world. The complete ban on plastic is not possible hence it is possible to reuse in construction industry as an aggregate [5], as an infill block for partition wall [6] etc., Other way to reuse plastic is through one of the new techniques is plastic cells filled concrete pavement block (PCFCPB) [7]. The cast in-situ concrete paving block technique was developed from 1994 to 2003 in South Africa by Vissar and Hall [8]. Also, the incorporation of construction and demolition waste from damaged structures to road development will reduce the ecological impact and solve dumping issues [9]. The low workability is needed for the concrete roads [10]. That will be fulfilled by the recycled concrete aggregate (RCA) from the C&D waste. The concrete with RCA gives similar results as conventional concrete properties such as strength, flexural strength, split tensile strength, abrasion resistance and rapid chloride permeability testing [11]. The waste stone dust is also preferred by some of the researchers to reduce the cost of construction [1].

Many researchers discussed about the benefits of concrete pavers such as availability in various size, colour and shape if having decorative landscapings are possible. The pre-casted pavers mostly used for parkings, walkways, facades of vertical walls, channel lining for water transportation through channels, protection of slopes and roads construction [12]. In a view to provide access in any climatic condition to all rural areas of country (around 6 Lakhs villages in India alone), the ministry of human resources, Government of India, introduced Pradhan Mantry Gram Sadak Yojna (PMGSY) in December, 2000. Still more than 50% of the rural villages are left to cover [13].

The rigid concrete pavement roads having better strength, longer life span and reduced cost of construction by introducing recycle, by-product and industrial waste. The plastic cell filled concrete paver block (PCFCPB) with addition of cementitious by-product materials gives better results [14]. The introduction of plastic cell filled concrete paver block technique will enhance technical properties act as a flexible-rigid pavement, high elastic modulus, reduce initial cost of construction and maintenance and ecological footprint [15]. The plastic cell filled concrete pavement block made with low density polyethylene (LDPE) of 200 micron to 0.49 mm thickness [12] and high-density polyethylene (HDPE) of thickness of 20–35 mm [16]. The size of each plastic cell filled concrete pavement block cell is 15 cm × 15 cm and depth may vary from 50 to 150 mm, most of research work considered 100 mm for their study [17]. The strips are welded by heat sealing or paddle sewing at an interval of 30 cm. Two different techniques of plastic cell filled pavers are shown in Figs. 1 and 2.

The compaction and curing of plastic cell filled concrete pavement block roads is similar to conventional concrete roads. The design thickness for flexible and rigid pavement is around 350–370 mm respectively in many layers but in the case of plastic cell filled concrete pavement block the thickness is reduced to 200 mm only in two layers. The costs of construction of flexible and rigid pavements are higher than the plastic cell filled concrete pavement block by about 9% and 150% respectively.

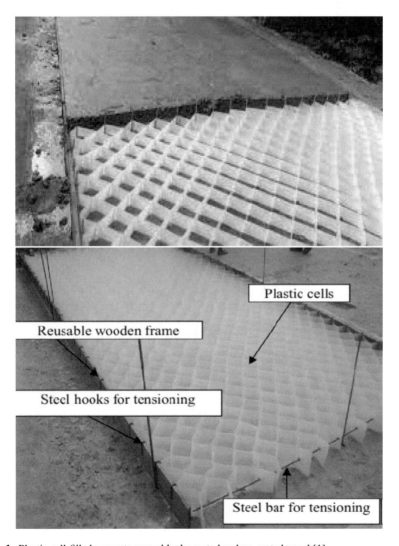

Fig. 1 Plastic cell-filled concrete paver blocks casted and un-casted panel [1]

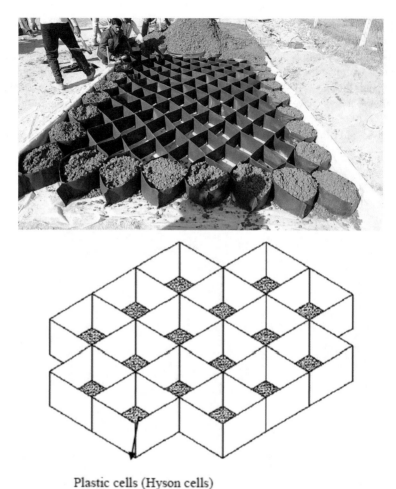

Plastic cells (Hyson cells)

Fig. 2 Plastic cell-filled concrete paver blocks during in-situ casting [1]

Similarly, the cost of maintenance in five years is also higher to about 43% and 141% than plastic cell filled concrete pavement block [1].

Now, the plastic cells are commercially available in market having a size of 0.20 mm to 0.35 mm. From the various previous research work on plastic cell filled concrete pavement block, the author concludes that use of cementitious material in the plastic cell filled concrete pavement block will give improved hardened properties, the study of mechanical properties also enhanced. The study on cost effectiveness is also carried out but till date very few studies available on ecological footprint assessment of plastic cell filled concrete pavement block (PCFCPB). The basic objective of the current study is to lower the ecological footprint of concrete road by adopting a new low-cost technique of PCFCPB using recycled raw materials.

2   Methods and Materials

2.1   Construction Materials

Construction of subgrade and subbase has been done as per Specifications for Rural Roads in India. The details of the construction materials are discussed as follows:

2.1.1   Plastic Cells

The formwork of plastic cells was made from recycled low-density polyethylene (LDPE) sheets of thickness 0.22–0.25 mm. Plastic sheet manufacturers were supplied with rolls of strips (100 mm wide) that are required for the pavement. The strips were stitched to form cells. The cells remain buried and the recycled LDPE sheets are usually rendered black in colour. Waste low density polyethylene (LDPE) is available in plenty and the recycled LDPE sheets of thickness 0.30 to 0.35 mm can be used for making the formwork of cells. Readymade formwork of cells also can be obtained from the market. A pair of strips can be also stitch at 300 mm interval. The photographical image of plastic cell is shown in Fig. 3.

2.1.2   Cement

Ordinary Portland cement (53 grades) was used for casting concrete for the sample pavement. The specific gravity of the cement is about 3.15.

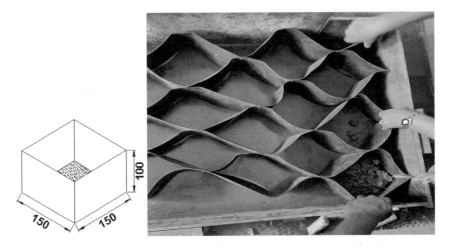

Fig. 3 Image of proposed plastic cell sample

2.1.3 Fine Aggregates

River sand was used to prepare the concrete for the sample pavement which is locally available in Malegaon city. The fine aggregate retained on 0.075 mm IS sieve. It should be clean screened dusts and should be free from organic matter, loam and clay.

2.1.4 Waste Building Materials

Waste building materials were used as a replacement of course aggregate in the sample of plastic cell-filled concrete pavement. Waste building materials are used in the sample pavement were consisting of waste concrete (M20 grade), wasted fire in clay bricks and waste cement mortar etc. The waste building materials are retained on 4.75 mm IS sieve size. The photographical images of the waste materials are used in the sample pavement are shown in Fig. 4.

The sample of the plastic cell-filled concrete pavement was prepared in the laboratory as per the Indian Standard. The sample of the proposed plastic cell-filled concrete pavement is shown in Fig. 5.

The third strip can be stitched/welded to the first pair at 300 mm interval so that the jointong lie at the centre of the previous stitching joints. The third and the fourth ones are again stitched/ welded like the first two. The quantity of plastic sheets of thickness 0.22 mm, side 150 mm and depth 100 mm may be nearly 2333 kg for a

Fig. 4 Waste concrete, mortar and bricks used in proposed pavement

Fig. 5 Plastic cell-filled concrete pavement

pavement of 3.75 m wide and 1 km long. Formwork of cells should be so stitched that one roll of cells for a pavement width of 3.75 m has a length of about 10 to 20 m upon stretching. The successive rolls of cell can be stapled for pavement construction. LDPE sheets may weigh more because of greater thickness requirement for stiffness.

2.2 Environmental Assessment

The study assesses environmental impact of plastic cell-filled concrete pavement by using Ecological Footprint indicator. The method of Ecological Footprint assessment of plastic cell-filled concrete pavement considered three parameters: (1) Material's environmental impact, (2) Machine's environmental impact and (3) labour's environmental impact. The details of environmental impact of plastic cell-filled concrete pavement are discussed as follows.

2.2.1 Material's Environmental Impact (EF$_{mei}$)

Materials of plastic cell-filled concrete pavement are accountable for substantial resource consumptions therefore; the Ecological Footprint of the plastic cell-filled concrete pavement should be examined. The EF$_{material}$ has been estimated by Eq. 1 [18].

$$\text{EFmaterial} \sum \left(\frac{\text{Cmi.Emi}}{\frac{\text{Af}}{1-\text{Aoc}}} \right) = .e_{CO_2 land} + \sum \left(\frac{\text{Cwi}}{\text{Ywi}} \right).e_i \qquad (1)$$

where,

C_{mi} is represented as material consumption of the ith material; E_{mi} is embodied emission of ith material. C_{wi} is consumption in the ith natural material, and Y_{wi} is materials productivity. A_f absorption factor of forests is considered to be 2.7 tCO_2/ha [19], A_{oc} (0.3 [20]) is the fraction of annual oceanic emission sequestration.

2.2.2 Machine's Environmental Impact ($EF_{machine}$)

The $EF_{machine}$ depends on machinery used for plastic cell-filled concrete pavement placement. The estimation of $EF_{machine}$ is given by Eq. 2.

$$EF_{machine} = \sum (E_i \cdot \alpha i) \left(\frac{1 - A_{oc}}{A_f} \right) \cdot e_{CO_2 land} \tag{2}$$

where E_i is the amount of energy/fuel consumed during the use of machineries; α_i is the emission factor of energy/fuel sources. The details of machinery impact are depicted in Table 1.

2.2.3 Labour's Environmental Impact (EF_{labour})

The EF_{labour} related the Ecological Footprint of food consumption by labour/manpower. The annual Ecological Footprint of food consumption per capita in India is reported in Table 2. EF_l is determined by Eq. 3.

$$EF_{labour} = FTE \cdot EF_{food} \tag{3}$$

where,

FTE is full time equivalence (i.e., 8 h/day working for 250 working in one year), EF_{food} represents the annual Ecological Footprint of food consumption by the labours.

2.3 Economic Assessment

The economic assessment of the asphalt pavement generally considers four parameters:

(1) Material cost,
(2) Machinery cost,
(3) Labour's cost,

In this study, transportation cost of materials, labourer and machineries have not been considered due to the large variation in data. The details of rest of the parameters are mentioned in Table 3.

Table 1 The details of machinery for pavement construction

Sr. No	Name of machinery required	Description of machine	Power capacity	Fuel/energy consumption
1	Paddle Sealer	Sealing speed 1–12 m/minute	600 W, 220 V	0.6 kWh/hr
2	Single drum concrete mixer	Engine Power:(HP at rpm) Operating weight:500 kg,	106 hp @ 2300 rpm	3.75–5 L diesel fuel working hour
3	Grader	Operating Weight 13,700 kg	Engine Power 160 hp	3.75–5 L diesel fuel working hour
4	Paver	Capacity: 2.50–9.5 m Brand/Make: POWER PAVER Model/Type: SF 1700 Hydraulic Cylinders: 4, Largest quantity of PQC paved in 24 h—14,613.30 m^3	Rated power at 2,100 rpm: 305 hp	224 kWh/120-to-150 m placement
5	Excavator	Brand: Hyundai, ISO 3046 Maximum Bucket Capacity: 0.65–0.72 m^3 HD SAE Heaped Max Digging Depth: 5500 mm	105 hp (78 kW) at 2200 rpm	33 L of diesel fuel per hour
6	Tiner/cure machine		–	13.2 L of diesel per hour
6	Needle Vibrator	Phase: Double Power: Petrol Weight: 35 kg Diameter: 25 × 40x50 × 60 mm^3 Material: Mild Steel Drive Mode: Engine Vibrator Phase Type: Single Automation Grade: Automatic Capacity: 3.5 hp	Power Consumption 1500 W,4800 rpm	Fuel Consumption 700 ml/hr
7	Plate Compactor	Capacity: 4.5 ton Engine Power: 110hp @ 2200 rpm, Engine Type: Petrol	Capacity 5 hp	3.73 kWh/hr

Table 2 Main raw materials for food preparation per capita in India

	Monthly consumption [21]	Total annual consumption	CO_2 emission factor	Yield Production [22, 23] (ton/ha)	EF (gha)
Cereals	9.28 kg	111.36 kg		2.39	0.117
Pulses	0.90 kg	10.8 kg		0.69	0.039
Vegetable	8.4 kg	100.8 kg		1.61	0.157
Beef	0.06 kg	0.72 kg		32	0.011
Mutton	0.08 kg	0.96 kg		72	0.006
Milk	5.4 L	64.8 L		458	0.062
Fish	0.252 kg	3.024 kg		0.035	0.030
Fruits	0.654 kg	7.848 kg		2330	1.58×10^{-6}
Edible oil	0.85 kg	10.2 kg		0.38	0.068
Wood	4.3 kg	51.6 kg	1.5–1.6 ($kgCO_2$/kg)	73 m³/ha	0.028
LPG	1.9 kg	22.8 kg	3.31 ($kgCO_2$/kg)		0.025
Kerosene	0.40 L	4.8 L	2.58 ($kgCO_2$/litre)		0.004
Total annual EF/person					0.549

Table 3 The details of parameters for economic assessment

Materials	#Cost (Rs)
Cement (OPC, 53 grade)	300/50 kg bag
Waste building materials	100/tonne
Fine aggregate	260/tonne
Plastic sheet (thickness 0.22–0.25 mm; 100 mm width)	60,000/tonne
Labour cost	500/day
Machinery	
Paver	2000/hr
Paddle sealer	500/day
Single drum concrete mixer	620/hr
Grader	1000/hr
Excavator	1500/hr
Tiner/cure machine	500/hr
Needle vibrator	500/day
Plate compactor	500/day

All the costs are taken from local market

3 Results

The study assesses the enviro-economic assessment of the plastic cell-filled concrete pavement for Indian conditions. The details of the Ecological Footprint assessment and Economic assessment of the plastic cell-filled concrete pavement construction are explained as follows.

3.1 Environmental Assessment

The Ecological Footprint of the plastic cell-filled concrete pavement construction (3.75 m width, 100 mm thickness and 1 km length) is estimated as 22–23.45 gha.

3.1.1 Material's Environmental Impact (EF_{mei})

The total volume of the concrete required to construct one km of the plastic cell-filled pavement is about 375 m^3. The Ecological Footprint of the M20 concrete (with waste building materials used as coarse aggregate) is 0.0428 gha/m^3. The Ecological Footprint of material consumption in the plastic cell-filled pavement construction is estimated as 17.42 gha. The Ecological Footprint of the materials are shown in Table 4.

Table 4 Ecological Footprint of material consumption per m^3 concrete

Materials	Unit	Quantity	Embodied Energy (MJ/unit)	Embodied emissions	Ecological footprint (gha/unit)
Cement	Kg	397.96	5.32 [24]	–	1.032×10^{-4}
Water	Kg	197	0.01kWh/m^3 (up to depth of 36.5 m) [25]		3.93×10^{-9}
Fine aggregate	Kg	663.27	464.29 [24]	–	0.012×10^{-5} [26]
Waste building materials	Kg	1193	–	–	0
Waste plastic strips	Kg	2.381		2.13 kgCO$_2$/kg [24]	0.0053
Water for Curing (for 28 days)	Litre	168	0.01kWh/m^3 (up to depth of 36.5 m) [25]		3.93×10^{-9}

Table 5 The environmental impact of machines used for pavement construction

Sr. No	Name of machinery required	Operational time (8 h = 1 working day)	Fuel/energy consumption	Ecological footprint (gha/hr)	Ecological footprint (gha)
1	Paddle Sealer	1–2 days	0.6 kWh per working hour	0.000163–0.000326	0.001305–0.0026
2	Single drum concrete mixer	2–3 days	3.75–5 L diesel fuel working hour	0.0034875–0.00465	0.0558–0.1116
3	Grader	2–3 days	3.75–5 L diesel fuel per working hour	0.00348–0.00465	0.1674–0.3348
4	Paver	6–8 days	224 kWh/120-to-150-m placement	0.06092	2.9245–3.8994
5	Excavator	4–5 days	33 L of diesel fuel per working hour	0.03069	0.98208–1.2276
6	Tiner/cure machine	28 h	13.2 L of diesel per working hour	0.012276	0.343728
7	Needle vibrator	48 h	Fuel Consumption 700 ml diesel per working hour	0.000651	0.031248
8	Plate compactor	48 h	3.73 kWh per working hour	0.001014	0.04869888

[#]Ecologic1al Footprint of grid electricity is about 2.72×10^{-4} gha/kWh[27]
[*]Ecological Footprint of diesel combustion is about 9.3×10^{-4} gha per liter [28]

3.1.2   Machine's Environmental Impact (EF$_{machine}$)

The environmental impact of the machine involves the construction of the plastic cell-filled concrete pavement mentioned in Table 5. The EF$_{machine}$ for the plastic cell-filled concrete pavement is 4.55–5 gha. The environmental impact of the machinery used in pavement construction is highest among all three parameters (i.e., materials, machinery use and labours).

3.1.3   Labour's Environmental Impact (EF$_{labour}$)

The estimated labour requirement for the proposed plastic cell-filled concrete pavement construction is 29–31.5 labour-days. The FTE for the pavement construction is 0.114–0.143. The Ecological Footprint of one labour -day is about 9.02×10^{-4} gha. The Ecological Footprint of labour's impact for the pavement construction has been estimated in the range of 0.0261–0.0329 gha. The environmental impact of the different parameters of pavement construction is depicted in Fig. 6.

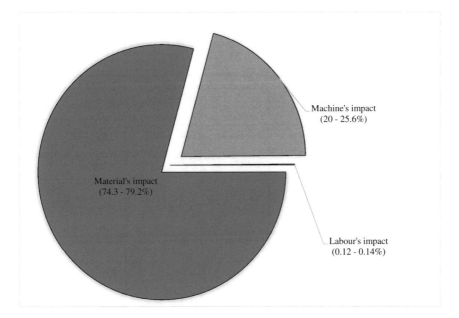

Fig. 6 The Ecological Footprint distribution of different parameters of pavement construction

3.2 Economic Assessment

The constructional cost of the 1-km plastic cell-filled concrete pavement (3.75 m width and 100 mm thickness) is estimated as Rs 1.47–1.53 million. The addition of the plastic cell may increase the pavement cost as compared to the conventional concrete pavement, however, waste building materials used as a replacement of the coarse aggregate may reduce the cost of the proposed pavement. The distribution of the pavement construction cost is depicted in Table 6.

4 Conclusions

Plastic cell-filled concrete pavement may be a solution for the rural roads (low traffic roads) because of the durability, and longer service life of the concrete pavement. Replacement of coarse aggregates by waste building materials is demonstrated to be cost cutting options without significant changes in the overall construction. Hence, the plastic cell-filled concrete pavement may be a suitable alternative to the conventional rigid pavement.

The Ecological Footprint of the plastic cell-filled concrete pavement construction (3.75 m width, 100 mm thickness and 1 km length) is estimated as 22–23.45 gha. The environmental impact of the materials consumed in the proposed pavement is the

Table 6 Economic assessment of the pavement construction

Sr. No	Name of machinery required	Operational time (8 h = 1 working day)	Quantity	Rate	Total cost (Rs)
1	Cement	–	149,235 kg	300 Rs/bag (50 kg each)	970,027.5
2	Water	–	73,875 L	1 Rs/kg	73,875
3	Fine Aggregate	–	248,726 kg	260/tonne	64,668.825
4	Waste building materials	–	447,375 kg	100/tonne	44,737.5
5	Waste plastic strips	–	2812.5 kg	60,000/tonne	1,269,881.3
6	Water for Curing	–	63,000 L	1 Rs/kg	63,000
7	Paddle Sealer	1–2 days	–	500/day	500–1000
8	Single Drum concrete mixer	2–3 days	–	620/hr	9920–14,880
9	Grader	2–3 days	–	1000/hr	16,000–24,000
10	Paver	6–8 days	–	2000/hr	96,000–128,000
11	Excavator	4–5 days	–	1500/hr	48,000–60,000
12	Tiner/cure machine	28 h	–	500/hr	14,000
13	Needle Vibrator	48 h	–	500/day	3000
14	Plate Compactor	48 h	–	500/day	3000
15	Labours	29–36.5 days	–	500/day	14,500–18,250
Total cost of plastic cell-filled concrete pavement (3.75 m width, 100 mm thickness and 1 km length)					1,474,801–1,536,011

highest (i.e., 74.3–79.2% of the total environmental impact of the proposed pavement) followed by the machinery environmental impact (20–25.6%) of the pavement. The constructional cost of the proposed plastic cell-filled concrete pavement is estimated as Rs 1.47–1.53 million. It's not significantly higher and it can be compensated with improved service life of the pavement. The life cycle cost assessment may also help to provide inclusive economics of the proposed pavement design. The proposed method of pavement construction may improve the sustainability of road construction by utilization of recyclable materials such as waste plastic sheets, E-waste and building waste materials.

References

1. Singh YA, Ryntathiang TL, Singh KD Economic evaluation of plastic filled concrete block pavement. Int J Eng Adv Technol (IJEAT) 5:57–65
2. Ismail ZZ, Al-Hashmi EA (2008) Use of waste plastic in concrete mixture as aggregate replacement. Waste Manag 28(11):2041–2047
3. Abid AYZ, Usama AA, Husain D, Sharma M, Prakash R (2022) Ecological footprint assessment of recycled asphalt pavement construction. In: Muthu SS (ed) Environmental footprints of recycled products. environmental footprints and eco-design of products and processes. Springer, Singapore. https://doi.org/10.1007/978-981-16-8426-5_5
4. Siddique R, Cachim P (2018) Waste and supplementary cementitious materials in concrete: characterisation, properties and applications. Woodhead Publishing
5. Silva RV, de Brito J (2018) Plastic wastes. In: Waste and supplementary cementitious materials in concrete. Woodhead Publishing, pp 199–227
6. Bhushaiah R, Mohammad S, Rao DS (2019) Study of plastic bricks Made from waste plastic. J Eng Technol 6(4):6
7. Ryntathiang TL, Mazumdar M, Pandey BB Suitability of cast in-situ concrete block pavement for low volume roads. In: 8th international conference on concrete block paving, November, pp 6–8
8. Visser AT, Hall S (1999) Flexible portland cement concrete pavement for low-volume roads. Transp Res Rec 1652(1):121–127
9. Verian KP, Warda A, Yizheng C (2018) Properties of recycled concrete aggregate and their influence in new concrete production. Resourc Conserv Recycl 133:30–49
10. Tattersall GH (1991) Workability and quality control of concrete. CRC Press
11. Manikandan T, Mohan M, Siddaharamaiah YM (2015) Strength study on replacement of coarse aggregate by reused aggregate on concrete. Int J Innov Sci Eng Technol 2:438–441
12. Sharma P, Batra RK (2016) Cement concrete paver blocks for rural roads. In: Futuristic trends in engineering, science, humanities, and technology FTESHT-16, p 70
13. Madke MP, Harle S (2016) Plastic cell filled concrete road: a review. J Struct Transp Stud 1(3)
14. Chkheiwer AH (2017) Improvement of concrete paving blocks properties by mineral additions. J Babylon Univ/Eng Sci 25(1):157–164
15. Singh YA, Ryntathiang TL, Singh KD (2012) Structural performance of plastic cell filled concrete block pavement for low volume roads. In: 10th International conference on concrete block paving Shanghai, Peoples Republic of China, pp 1–25
16. Johnson S, Faiz A, Visser A (2019) Concrete pavements for climate resilient low-volume roads in pacific Island countries
17. Singh YA, Ryntathiang TL, Singh KD (2012) Structural performance of plastic cell filled concrete block pavement for low volume roads. In: 10th International conference on concrete block paving Shanghai, Peoples Republic of China, pp 1–25
18. Husain D, Prakash R (2019) Ecological footprint reduction of built envelope in India. J Build Eng 21:278–286. https://doi.org/10.1016/j.jobe.2018.10.018
19. Mancini MS, Galli A, Niccolucci V, Lin D, Bastianoni S, Wackernagel M, Marchettini N (2016) Ecological footprint: refining the carbon footprint calculation. Ecol Indic 61(2):90–403
20. Scripps Institution of Oceanography (SIO) (2017) "The keeling curve". https://scripps.ucsd.edu/programs/keelingcurve/2013/07/03/how-much-co2-can-the-oceans-take-up/. Accessed 23 Mar 2020
21. National sample survey office (NSSO) (2014) Household consumption of various goods and services in India 20ll-12, ministry of statistics and programme implementation, government of India, June 2014 mospi.nic.in/mospi_new/upload/Report_no558_rou68_30june14.pdf
22. Chambers N, Simmons C, Wackernagel M (2004) Sharing nature's interest: ecological footprints as an indicator of sustainability. Sterling Earthscan, London, Great Britain
23. Indian Horticulture Database (IHD) (2014) Ministry of agriculture, government of India. www.nhb.gov.in

24. Inventory of Carbon & Energy (ICE). Sustainable Energy Research Team (SERT) (2011). www.bath.ac.uk/mech-eng/sert/embodied. Accessed Sept 2022
25. Plappally AK, Lienhard JHV (2012) Energy requirements for water production, treatment, end use, reclamation, and disposal. Renew Sustain Energy Rev 16(7):4818–4848. ISSN 1364-0321. https://doi.org/10.1016/j.rser.2012.05.022
26. Husain D, Prakash R (2019) Life cycle ecological footprint assessment of an academic building. J Inst Eng (India) Ser A 100(1):97–110. https://doi.org/10.1007/s40030-018-0334-3
27. Biswas A, Husain D, Prakash R (2021) Life-cycle ecological footprint assessment of grid-connected rooftop solar PV system. Int J Sustain Eng 14(3):529–538. https://doi.org/10.1080/19397038.2020.1783719
28. Husain D, Ahmad A, Sharma M, Tewari K, Prakash R (2022) Ecological footprint of jatropha biodiesel production at low scale. In: Environment & energy, environmental science, engineering and technology, environmental sciences, Newly Published Books, Nova, Science and Technology, Special Topics. ISBN: 978-1-68507-548-4. https://doi.org/10.52305/KJOY6010

Carbon Footprint and Economic Assessment of LED Bulbs Recycling

Mohammed Salman Baig, Dilawar Husain, Shameem Ahmad, Fahad Bilal, Faheem Ansari, Sajid Naeem, and Manish Sharma

Abstract The light-emitting diodes (LED) in recent times, have become the most energy-efficient and rapidly developing lighting technology. However, it has been found that LED contains various perilous valuable materials that make their recycling essential. It also helps to recover rare earth elements and maintains the eco-balance of the planet. The environmental and economic assessments are being carried out in this study. The CO_2 emissions for recycling one tonne of LED bulbs were estimated to be in the range of 6.21×10^{-3}–5.52×10^{-2} tCO_2. The cost of LED bulbs recycling is estimated as Rs 20,221–20715 per tonne. The LED recycling should be promoted to reduce the cost of material processing (such as Ag, Au, As, Ga, Cu, Fe, Zn and other rare earth metals etc.), However, recycling comes at a cost of energy, process operations and manpower resources. LED bulb recycling is cost effective, eco-friendly and also helps to achieve the United Nations Sustainable Developmental Goals.

Keywords LED · Recycling · Carbon footprint · Economic assessment · Environmental assessment

M. S. Baig · S. Ahmad · F. Bilal · F. Ansari
Department of Electronics & Telecommunication Engineering, Maulana Mukhtar Ahmad Nadvi Technical Campus, Malegaon 423203, India

D. Husain
Department of Mechanical Engineering, Maulana Mukhtar Ahmad Nadvi Technical Campus, Malegaon 423203, India

S. Naeem
Department of Applied Sciences, Maulana Mukhtar Ahmad Nadvi Technical Campus Malegaon, Malegaon 423203, India

M. Sharma (✉)
Department of Mechanical Engineering, Malla Reddy Engineering College, Hyderabad, Telangana 500100, India
e-mail: manish.mvs@gmail.com

© The Author(s), under exclusive license to Springer Nature Singapore Pte Ltd. 2023
S. S. Muthu (ed.), *Environmental Assessment of Recycled Waste*,
Environmental Footprints and Eco-design of Products and Processes,
https://doi.org/10.1007/978-981-19-8323-8_3

1 Introduction

Rapid population growth and urbanisation has led to an unprecedented increase in energy, raw material consumption. Consequently, the waste generation rates are alarmingly increasing that threaten to contaminate the ecosystem. The World Bank predicted that between 2016 and 2050, the amount of waste generated worldwide would rise by 70% [1]. Currently, the trajectory for global waste generation shows that it will grow from 2.24 billion tonnes in 2020 to nearly 3.88 billion tonnes by 2050 [1].

However, e-waste creation worldwide is close to 53.6 million tonnes, of about 2.39% of all garbage produced globally. E-waste produced by electronic equipment contains several extremely valuable materials, including palladium, rare earth elements, nickel, copper, indium, gold, and others [2, 3]. Moreover, it also contains some dangerous materials like lead, mercury, brominated flame retardants, etc., which, if not handled appropriately, can have an adverse effect on the ecosystem and human health [2, 3]. According to [2, 3], Asia is the region with the biggest production of e-waste, with 40% of that garbage produced in China, 11.5% in Japan, and roughly 11% in India. In recent years, conventional light sources like incandescent bulbs and compact fluorescent lamps (CFLs) have frequently been replaced by light-emitting diode (LED) lamps, which are incredibly energy-efficient and ecologically beneficial. It is estimated that LED lamps' market share are expected to be about 75% of the lighting market by 2030 [4]. An LED's estimated lifespan is typically between five and seven years, and its end of life span is a concern as it generates a substantial amount of e-waste and if they are not treated properly can have a dangerous impact on the entire eco-system [5]. Figure 1 demonstrates the penetration rate of LED lamps worldwide. The LED lighting market is growing at an annual growth rate of approximately 14.25% over the next 5 years. Increased usage of LED lamps creates a growing concern of their end-of-life disposal. This adds to the current ongoing challenge of LED bulbs recycling. LED recycling is important as it contains substantial amounts of valuable metals such as gold (Au), arsenic (As), gallium (Ga), indium (In), and rare-earth elements (REEs) such as (yttrium, europium, gadolinium and cerium etc.) in phosphor powder [6].

Different methods are employed in literature to recycle LED bulbs. In [8], Junbeum Kim applied lessons from the circular economy to the recycling of LED bulbs and offered potential solutions. In order to recycle cerium, europium, and yttrium from LED flat panel displays, in his notional chemical process, Ruiz-Mercado [9] included various processes for leaching, solvent extraction, ion exchange, precipitation, chemical reaction, and calcination. Critical metals in discarded LEDs are extracted using recycling techniques including hydrometallurgy, mechanochemical activation, vacuum metallurgy, etc. One of the significant metals that is typically present in LED chips is the rare element gallium. They often take the form of compounds like indium gallium nitride (InGaN), gallium nitride (GaN), and gallium arsenide (GaAs), which are employed in a variety of optoelectronic devices (LEDs etc.). According to a research, the market for GaN devices will reach around $600

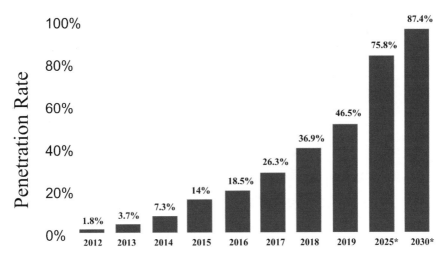

Fig. 1 Global penetration rate of LED lamps [7]

million by 2020, and the annual growth rate of the GaN compound is predicted to be over 80% through that year [10]. As a result, Ga from used LEDs was recycled. Seyyed Mohammad Mousavi [11] examined the bio-hydrometallurgy-based extraction of Cu, Ni, and Ga from waste LEDs. The modified A. ferrooxidans leached about 84% copper, 96% nickel, and 60% gallium, based on the results.

In general, the recycling process, as depicted in Fig. 2, depends on the resources' availability and viability from an economic standpoint. Keeping these ideas in mind, the stockpiling of basic materials and their aggregation into a single fraction are important steps in the development of LED recycling systems. Without a doubt, the lengthy lamp lifetimes and the resulting low return rate of waste LED lamps (1% for Germany in 2016) make the requirement for specific recycling methods for LED bulbs unnecessary [12]. Nonetheless, the dominance and augmentation of the total lighting market by LED innovation obviously focus on the point of view that the appropriate treatment of LED squander streams will become important soon. To foster proper reusing advancements for LED-based lighting hardware today and to explore fitting division and extraction strategies for the LED-explicit important components permits to be proactive instead of receptive. Besides, as to the low garbage removal costs that are paid for LED lighting gear laying out a different assortment and reusing framework for LED lights and gas release lights is viewed as fitting.

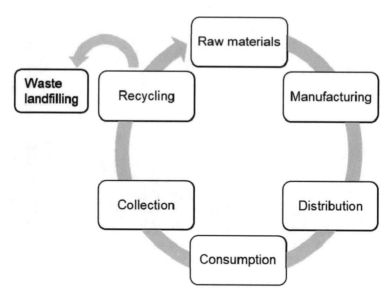

Fig. 2 Conceptual diagram of the circular economy

1.1 An Economic LED Recycling Approach

Figure 3 shows a schematic of the recycling process required for separating the component and the material fractions. The crucial step is extracting the coarse fragmentation. Using modified procedures, the process is sorted and categorised after the resulting mixed fraction. To separate the metals that may be magnetised, for instance, metal separators are used. Utilizing flotation techniques, materials with strong densities, such as plastic and ceramic, are separated. Sieving is a technique used to separate materials with various grain sizes. To remove copper from electromagnetic coils, the collected electronic components are processed through e-waste recyclers.

The LED packages resulting from the main fractions are at first treated as impurities, and due to their intense fluorescence, they can be easily detected by exposing them to UV-light irradiation. Currently, there are no standard procedures to recover the critical elements from LED. However, waste phosphors from fluorescent lamps are further collected and stored using commonly used procedures.

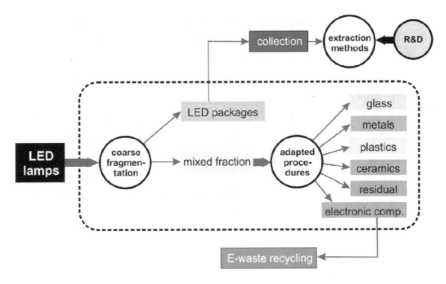

Fig. 3 General recycling processes of LED bulbs

2 Methods and Materials

2.1 LED Bulbs Characterisation

An optoelectronic device called an LED bulb uses the electro-luminance principle to turn electrical energy into light energy. A strong electric field causes the semiconductor material used in LED lights to emit light. LED bulbs frequently use semiconductor materials like gallium arsenide (GaAs), gallium phosphide (GaP), and gallium arsenide phosphide (GaAsP), among others. The colour of the radiated light depends on the semiconductor material. The colour details are shown in Table 1.

The components found in the LED bulbs are given as:

Table 1 The colour of radiated light depends on semiconductor materials

Semiconductor Materials	Colour	Forward Voltage (In Volts)
GaN	White	4.1
GaN	Blue	5.0
GaP	Green/Red	2.2
GaAsP	Yellow	2.2
GaAsP	Red	1.8
AllnGaP	Yellow	2.1

Polymeric Housing and Polymeric Cover: A combination of polyethylene and poly-butylene terephthalate makes up the polymeric housing (widely used as insulation). Polycarbonate is used to make the polymeric cover (PC).

Metallic Housing: This consists of composed aluminium and is dented after the comminution process.

Bulb base: Moulds are used in the bulb base's construction. They have a screw-shaped depression so they may easily fit into the socket of different light fittings.

Polychlorinated Biphenyls (PCB): A typical polychlorinated biphenyl is made up of several different kinds of metals, including expensive metals (such as platinum, gold, and silver etc.), base metals (such as steel, aluminium, and copper etc.), and toxic substances (lead, arsenic, antimony, and mercury etc.).

2.2 Recycling Process

In this section, the mechanical process of LED recycling has been considered in this part. After LED bulbs reach their End of Life (EoL), the recycling process involves a number of mechanical processing steps, including (a) comminution process, (b) mechanical sieving process, (c) magnetic separation process through two conveyor belts, (d) electrostatic separation process, and (e) gravity separation. Numerous components of the LED bulbs were successfully separated to a great extent, thanks to these processes. The flow design for recycling LED bulbs is shown in Fig. 4.

(a) **Comminution**
    For high size reduction and straightforward manipulation of the particle size range, hammer milling is frequently employed for the process of LED comminution. Typically, a hammer mill's grinding energy ranges from 5 to 60 kWh/tonne [13].

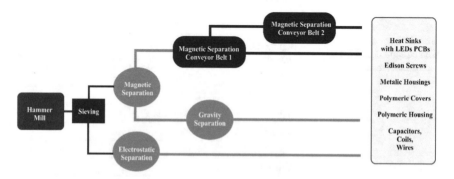

Fig. 4 Steps of LED bulb recycling

(b) **Sieving**

Mechanical vibrating shaker is generally used to separate the larger size materials from the particulates (product of comminution process). The power capacity of the sieving machine varies between 1 and 2 hp and a sieving capacity of 2–4 tonne per hour for 20–25 mm size particles. As per Fig. 4, the retained materials of the sieving process were passed to the magnetic separation process while the passing materials of the sieving process were routed to the electrostatic separation process.

(c) **Electrostatic Separation**

The passing materials of the sieving process further move to the electrostatic separation process for split of conductive materials and non-conductive materials. The general energy consumption in electrostatic seperation varies between 0.2 and 1 kWh per tonne of material seperation [14].

(d) **Magnetic Separation**

Magnetic separation is the process of removing the ferromagnetic material from the sifting of retaining materials. From the retained portion of the sieving, magnetic separators are utilised to separate the materials with higher magnetic vulnerability. Standard magnetic separation units ensure a separation process that is continuously carried on a moving stream of dry/wet particles passing through low or high magnetic fields. Drum, cross-belt, roll, high-gradient magnetic separation (HGMS), high-intensity magnetic separation (HIMS), and low-intensity magnetic separation (LIMS) types of magnetic separators are frequently employed. Figs. 5 and 6 illustrate the two basic designs of magnetic separators. A non-magnetic drum is fitted with three to six permanent magnets in a Drum type magnetic separator as shown in Fig. 5. The inner periphery of the Drum type is composed of ceramic or rare earth magnetic alloys. A moving stream of wet feed is passed through the rotating drum at a uniform motion. The rotating magnets of the drum collect the ferromagnetic minerals that are passed on to the outer surface of the drum. The general capacity of magnetic separator varies between 3.7 kW to 10.62 kW with a separation capacity of 1–2 tonne per hour [15]. A Cross-belt magnetic separator shown in Fig. 6, consists of a magnet mounted over the moving belt that carries the magnetic feed.

(e) **Gravity Separation**

Materials of different specific gravities are easily separated using simple gravity separators. The process is carried out by the materials' motion in response to the specific gravity, together with other external forces such as air or water adding resistance to motion. The size, shape, and specific gravity of the moving material affect its motion in a fluid. When there are slimes present, the efficiency decreases but becomes more effective as the size gets coarser. Gravity separators come in a variety of designs that are appropriate for various uses. The non-magnetic portion, primarily made up of the metal housing and polymer cover, was separated by gravity separation using a dense medium. The mass related energy consumption of 0.2 to 8 kWh/tonne and specific mass flow rates of 3 to 16 tonne/(m2-h) were obtained as in [16].

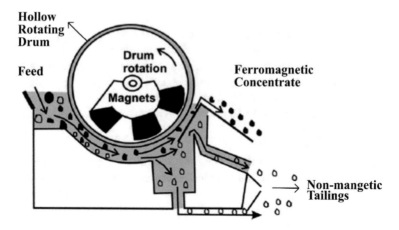

Fig. 5 Drum type magnetic separator [15]

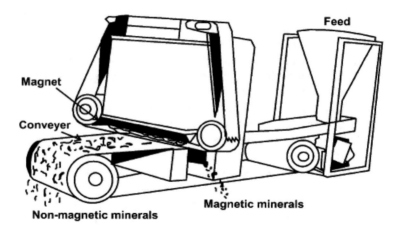

Fig. 6 Cross-belt magnetic separator [15]

2.3 Carbon Footprint

Carbon Footprint estimates greenhouse gas (GHG) emissions' concentration. All emissions (direct and indirect) are interpreted into an "intensity" number by expressing it in relation to the basic characteristics of an industry. Intensity, therefore, can be used to assess an industry as it raises over a period (gradually and/or via attainments). Intensity also helps the comparison of industries of different production capacities (magnitudes) and industries in different types of productions [17]. Carbon Footprint assessment is one aspect/metric of environmental impact investigation technique that can indicate where an industry/product or an industrial complex is most exposed to emission levels (i.e., above or below the sustainable emission benchmark).

3 Results

LED recycling involves a number of processing steps, including sieving process, magnetic separation process, electrostatic separation process, and gravity separation process. During the recycling process, all parts of the LED bulbs undergo a high degree of separation incurring losses. Figure 7 shows the mass balance for recycling one tonne LED bulbs. In a recycling plant, the LED bulbs are first crunched by the hammer mill (for small pieces) which incurs a loss of 31.1 kg (i.e., 3.11% of the 1 tonne LED bulb waste). The comminuted bulbs are then passed through a sieving that incurs a loss of 8.6 kg (i.e., 0.86% of the 1 tonne LED bulbs waste). The sieved materials are then sent to an electrostatic separator for split of conductive materials and non-conductive materials. The retained materials (i.e., 618.9 kg) of the sieving are passed through the magnetic separator to recover ferromagnetic metals. The remaining magnetic fraction (i.e., 248.6 kg) is further processed through two stages of magnetic separation with two different conveyer belt systems that yield the bulb base (i.e., 59.6 kg) and PCBs (i.e., 74.9 kg). The magnetic fraction (i.e., 223.2 kg) resulting from the conveyer belt system II of the magnetic separator is further processed yielding heat sinks with LEDs (i.e., 113.9 kg). The non-magnetic fraction consisting of the metal housing (i.e., 139.9 kg) and polymer cover (i.e., 230.2 kg), are separately recovered by gravity separation. The details of Carbon Footprint of different processes involved in LED bulb recycling are shown in Table 2.

Carbon Footprint of recycling of one tonne of LED bulbs has been assessed in this study. All the direct or indirect CO_2 emissions during each step of the recycling process have been accumulated to calculate the CF of LED recycling. The estimated CO_2 emissions are in the range of 6.21×10^{-3}–5.52×10^{-2} tCO_2 for the recycling (through mechanical process) of one tonne of LED bulbs. The Carbon Footprint distributions are shown in Fig. 8. It is seen that maximum emissions are involved in the comminution process (i.e., 65.9–89.2% of the total emissions) followed by the magnetic separation process (i.e., 4.8–29.8% of the total emissions) of LED recycling.

3.1 *Economic Assessment*

The cost of LED recycling facility consists of components such as waste LED cost, operation & maintenance cost of recycling plant etc. For estimation of LED recycling cost, input materials/resources such as waste LED bulbs and electricity cost (Rs 7.5/kWh in India) have been considered from the local market. The maintenance cost of the recycling plant has not been considered in this study. The labour requirements

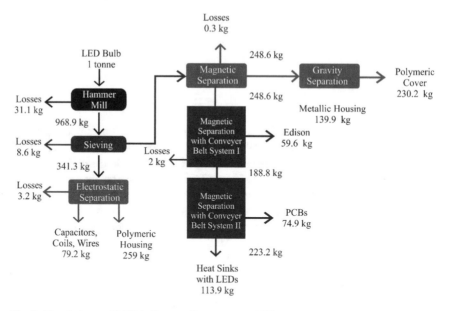

Fig. 7 Mass balance of LED bulbs recycling processes [18]

Table 2 Carbon Footprint of different processes

Process	Machine	Capacity	Power capacity or energy consumption	Carbon footprint (tCO$_2$/tonne)
Comminution	Hammer mill	–	5–60 kWh/tonne	0.0041 to 0.0492
Sieving	Horizontal vibrating sieve machine	2–4 tonne per hour	1–2 hp	1.5×10^{-4} to 6.1×10^{-4}
Electrostatic separation	Electrostatic separator	–	0.2 -1 kWh/tonne	1.64×10^{-4} to 8.2×10^{-4}
Magnetic separation	Magnetic separator	1–2 tonne per hour	3.7–10.62 kW	0.003 to 0.0043
Gravity separation	Water flow gravity separator	–	0.2 to 8 kWh/tonne	1.64×10^{-4} to 6.56×10^{-3}

[#] Emission factor of electricity generation in India is round0.82 tCO$_2$/MWh [19]

for recycling are estimated as 1.17 Full Time Equivalent(1FTE = 8 h of working per day for one year or 250 days) for 1000 tonne of E-waste recycling [20]. The details of the LED recycling cost are shown in Table 2. The cost of LED bulbs recycling is estimated (approximation) as Rs 20,221–20715 per tonne.

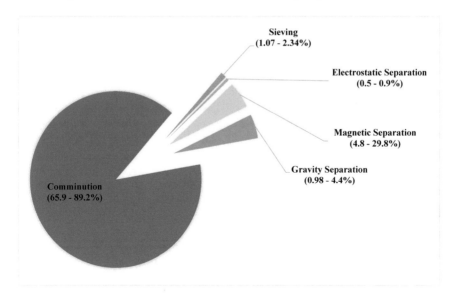

Fig. 8 Carbon Footprint distribution of LED bulbs recycling processes

Table 3 Economic assessment of LED recycling

Process	Power capacity or energy consumption	Cost (Rs/tonne)
Waste LED bulbs	–	20,000
Labour (@ 500 Rs/day)	–	146.3
Recycling processes		
Comminution	5–60 kWh/tonne	37.5–450
Sieving	1–2 hp	5.5–11.2
Electrostatic separation	0.2–1 kWh/tonne	1.5–7.5
Magnetic separation	3.7–10.62 kW	27.8–39.8
Gravity separation	0.2 to 8 kWh/tonne	1.5–60
Total recycling cost		~ **20,221–20,715**

* All the data taken from the local market

4 Conclusions

This research work evaluates the carbon footprint of recycling one tonne of LED bulbs. LED bulbs recycling involves a series of operations such as the comminution process, sieving process, magnetic separation process, electrostatic separation process, and gravity separation process. The CO_2 emissions for recycling one tonne of LED bulbs were estimated to be in the range of 6.21×10^{-3}–5.52×10^{-2} tCO_2. It is seen that maximum emissions are involved in the comminution process (i.e., 65.9–89.2% of the total CO_2 emissions) followed by the magnetic separation process ((i.e., 4.8–29.8% of the total emissions)) of LED recycling.

Results have indicated that there is lot of scope and potential for improvement in recycling of end-of-life LED bulbs. Further research may be required to improve the recovery rate, using more eco-friendly materials for the production of efficient, economical and sustainable LED bulbs.

References

1. Kaza S, Yao LC, Bhada-Tata P, Van Woerden F (2018) What a waste 2.0: a global snapshot of solid waste management to 2050. World Bank, Washington, DC
2. Biswas A, Husain D, Prakash R (2021) Life-cycle ecological footprint assessment of grid-connected rooftop solar PV system. Int J Sustain Eng 14(3):529–538. https://doi.org/10.1080/19397038.2020.1783719
3. Husain D, Tewari K, Sharma M, Ahmad A, Prakash R (2022). Ecological footprint of multi-silicon photovoltaic module recycling. In: Muthu SS (ed) Environmental footprints of recycled products. Environmental footprints and eco-design of products and processes. Springer, Singapore. https://doi.org/10.1007/978-981-16-8426-5_3
4. Nikulski JS, Ritthoff M, von Gries N (2021) The potential and limitations of critical raw material recycling: the case of LED lamps. Resources 10(4):37. https://doi.org/10.3390/resources10040037
5. Mizanur Rahman SM, Kim J, Lerondel G, Bouzidi Y, Nomenyo K, Clerget L (2017) Missing research focus in end-of-life management of light-emitting diode (LED) lamps. Resour Conserv Recycl 127:256–258. https://doi.org/10.1016/j.resconrec.2017.04.013
6. Bessho M, Shimizu K Latest trends in LED lighting. IEEJ Trans Fundam Mater 131(1):315–320. https://doi.org/10.1541/ieejfms.131.315
7. Statista (2021). LED penetration rate of the global lighting market based on sales from 2012 to 2030. https://www.statista.com/statistics/246030/estimated-led-penetration-of-the-global-lighting-market/
8. Rahman SM, Mizanur JK, Gilles L, Youcef B, Laure C (2019) Value retention options in circular economy: issues and challenges of LED lamp preprocessing. Sustainability 11(17):4723. https://doi.org/10.3390/su11174723
9. Ruiz-Mercado GJ, Gonzalez MA, Smith RL, Meyer DE (2017) A conceptual chemical process for the recycling of Ce, Eu, and Y from LED flat panel displays. Resour Conserv Recycl 126:42–49
10. Global Power GaN Market 2014: Forecasts to 2020. http://www.researchandmarkets.com/research/hphw5b/power_gan_market
11. Pourhossein F, Mousavi SM (2018) Enhancement of copper, nickel, and gallium recovery from LED waste by adaptation of Acidithiobacillus ferrooxidans. Waste Manag 79:98–108. https://doi.org/10.1016/j.wasman.2018.07.010
12. Recent data from the German National Register for Waste Electric Equipment ("Stiftung Elektro-Altgeräte Register ear") 2016
13. Tumuluru JS, Heikkila DJ (2019) Biomass grinding process optimization using response surface methodology and a hybrid genetic algorithm. Bioengineering 6:12
14. Magna Tronix. https://www.magnatronix.in/search.html?ss=electrostatic.
15. Magnetic Separators. http://www.kanetec.co.jp/en/pdf/120_138.pdf

16. Tomas J (2004) Gravity separation of particulate solids in turbulent fluid flow. Part Sci Technol 22(2):169–187. https://doi.org/10.1080/02726350490457222
17. IPCC. https://www.ipcc.ch/publications_and_data/ar4/wg1/en/spmsspm-projections-of.html
18. Martins TR, Tanabe EH, Bertuol DA (2020) Innovative method for the recycling of end-of-life LED bulbs by mechanical processing. Resour Conserv Recycl 161:104875
19. Ministry of Power Central Electricity Authority, Government of India (MPCEA) (2021) CO_2 baseline database for the indian power sector, user guide 2016" Accessed Nov 2021
20. United States Environmental Protection Agency (2020) Recycling economic information report. https://www.epa.gov/smm/recycling-economic-information-rei-report#findings

An Empirical Investigation of Waste Management and Ecological Footprints in OECD Countries

Bekir Çelik⊕, Doğan Barak⊕, and Emrah Koçak⊕

Abstract One of the most important issues addressed by the United Nations to realize sustainable development is environmental pollution. Even though factors involved in environmental degradation have been extensively studied and discussed in the literature, it lacks incorporation of the impact of waste management on environmental degradation. This study aims to fill the mentioned gaps by investigating the impact of waste management on environmental degradation. We use municipal waste per capita, the rate of municipal waste sent to landfills, recycling rate, per capita income, renewable energy consumption, trade openness, and ecological footprint per capita from 1995 to 2018 in OECD countries data. Fully modified ordinary least squares (FMOLS) and Pooled Mean Group (PMG) estimation tests are employed to test this relationship. According to the findings, a long-term relationship has been found. The long-term coefficient shows that the amount of municipal waste per capita increases the ecological footprint per capita. On the other hand, the amount of municipal waste sent to landfills, recycling rate, and renewable energy consumption reduce the ecological footprint per capita. Additionally, the Environmental Kuznets Curve (EKC) is valid for OECD countries. Good waste management has been playing an important role in reducing environmental degradation and these results can be very important guidance for policymakers.

Keywords Ecological footprint · Municipal waste · Landfill · Recycling · Renewable energy consumption · EKC

B. Çelik (✉)
Department of Economics, Nuh Naci Yazgan University, Kocasinan-Kayseri 38170, Turkey
e-mail: bcelik@nny.edu.tr

D. Barak
Department of Economics, Bingöl University, Selahaddin-i Eyyubi-Bingöl, Bingöl 12000, Turkey
e-mail: dbarak@bingol.edu.tr

E. Koçak
Department of Economics, Erciyes University, Melikgazi, Kayseri 38039, Turkey
e-mail: emrahkocak@erciyes.edu.tr

© The Author(s), under exclusive license to Springer Nature Singapore Pte Ltd. 2023   43
S. S. Muthu (ed.), *Environmental Assessment of Recycled Waste*,
Environmental Footprints and Eco-design of Products and Processes,
https://doi.org/10.1007/978-981-19-8323-8_4

1   Introduction

Sustainable Development Goal (SDG) policy is among the most important issues that the United Nations (UN) prioritizes. The UN has studies on 17 main topics and 169 sub-topics to achieve sustainable development goals [87, 152]. The World Bank defines municipal solid waste as "non-hazardous wastes from households, commercial and workplaces, institutions and non-hazardous industrial process wastes, agricultural wastes and sewage sludges". The waste problem is an important issue that all countries have to deal with [34, 52, 76]. With the rapid population growth, urbanization, and the increase in people's consumption habits, waste generation is also increasing [70, 80, 103, 125]. In the period from 1980 to 2018, the amount of solid waste for OECD countries increased from 395 million tons to 680 million tons [105]. Likewise, according to estimates using 2020 data, it is thought that one person produces 2.24 billion tons of solid waste, which corresponds to a 0.79-kg footprint. By 2050, it is thought that the amount of waste produced will be 73% higher than that estimated in 2020 [82, 148].

Every non-recyclable waste has to be collected in one area. These storage areas are generally the pits formed after the working areas of the sand quarries. The storage of waste is a method widely used all over the world [153]. Solid waste management, consists of gathering the waste generally via municipal, recycling the waste that collects from the house, commercial, and workplaces, and landfilling the remaining waste. Solid waste management is directly linked with SDG 8.4. "Improve progressively, through 2030, global resource efficiency in consumption and production and endeavor to decouple economic growth from environmental degradation, by the 10-year framework of programs on sustainable consumption and production, with developed countries taking the lead", SDG 12.2. "By 2030, achieve the sustainable management and efficient use of natural resources" and SDG 14.1. "By 2025, prevent and significantly reduce marine pollution of all kinds, in particular from land-based activities, including marine debris and nutrient pollution" [12, 105, 152].

Waste does not always result in pollution, but if it does result in pollution, the harm to the environment is quite high [22]. According to the sustainable development program published by the United Nations, the issue of environmental pollution is among the issues that are considered to be of high importance. When compared to developed countries, developing countries and less developed countries are faced with the waste problem more intensely [70]. Especially in underdeveloped countries, 90% of the waste is in the form of unregulated landfills, and the burning of garbage causes serious damage to individuals in terms of both the environment and health [70, 76, 148, 150]. The share of the greenhouse gas generated by the solid wastes of the consumers in the greenhouse gas emissions generated in the world is approximately 5% (1,460 $mtCO_2$ e) [70]. Similarly, it was concluded that solid waste generation contributes between 8 and 12% to greenhouse gas emission rates [143].

Especially in recent years, the focus on global warming and environmental pollution has increased the importance of this issue even more. The collection of solid wastes, the separation, and recycling of collected wastes, and the landfill of solid

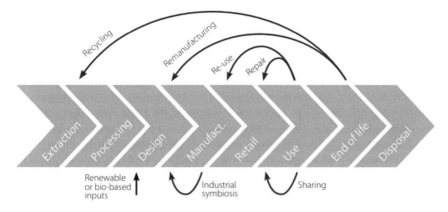

Fig. 1 Circular economy process diagram. *Source* OECD [106]

wastes are directly related to the environment [82]. After solid waste collection, the recycling process begins. Recycling is the process of collecting and processing the products to be thrown in a certain area and turning them into a product again. Every recycled product will provide a significant gain for society and the environment. The recycling of products can contribute to reducing the amount of waste generation, protecting natural resources, ensuring economic security via the protection of domestic products, reducing raw materials' need, saving energy, supporting production, and finally providing employment for new people to work in this field [28, 52, 56]. According to the European Environment Agency (EEA) report, by recycling glass, paper, batteries, motor oil, and aluminum cans, the amount of waste burned can be reduced by approximately 70%. Recycling is part of the circular economy program [26, 53, 163]. This situation is seen in the image of the circular economy model in Fig. 1.

According to the Environmental Protection Agency (EPA) waste management 2022 report made by data from Toxics Release Inventory (TRI), an additional 5 billion pounds of revenue was generated in the period 2011–2020 by waste storage and increased recycling. As it is known, very intense greenhouse gas emissions occur during the process of obtaining resources. For this reason, it is important to develop a recycling policy to reduce environmental pollution in the process of obtaining resources. Therefore, the negative pressure on the environment will be reduced by recycling [26, 57].

Storing solid wastes in landfills is the key issue for sustainable development and the environment. The landfilling of solid waste is a subject of close interest to all countries of the world. However, this creates a significant cost for countries [70]. Since 2000, the World Bank has provided nearly $5 billion in incentives to nearly 350 solid waste management programs in various parts of the world [148]. For this reason, the number of landfills in the world is increasing day by day [76].

Based on all this information, it has been observed that solid waste management is among the issues of high importance in sustainable development and environmental protection. Although there are a lot of studies about environmental pollution in the literature [88, 89], the number of empirically examine the relationship between solid waste management and environmental pollution studies is extremely limited. In this context, the relationship between solid waste management and environmental pollution will be investigated in our study. Additionally, testing the validity of environmental pollution, renewable energy consumption, trade openness, and the EKC hypothesis is another aim of the study. In this respect, it is thought to fill an important gap in the literature. We employed FMOLS and PMG methods to test the relationship between solid waste management and environmental pollution.

The study is structured into five parts; the first part is the introduction. The second part is associated with the literature review. The third part explains the data and methodological framework. The fourth part addresses the empirical findings of the analysis, while the last part concludes the study and provides some important policy recommendations.

2 Literature Review

When the literature on the study is examined, it has been observed that the number of studies that directly deal with the relationship between solid waste management and environmental pollution is quite limited. In the literature, firstly, the relationship between solid waste management and environmental pollution, which is directly related to the study, is discussed. Then, EKC hypothesis studies and analyzing the relationship between energy consumption and environmental studies are included, respectively. Lastly, investigated studies dealing with the nexus between other economic and social factors and environmental degradation are examined.

Eleftherios [55] discussed the connection between recycling and air pollution with Massachusetts data from the year 2009 to 2012. The panel fixed affect regression method has been used for empirical analysis. Also, the EKC hypothesis was checked for validity for recycling. An inverted U-shaped EKC has been found and that means air pollution and GDP have a negative correlation till the turning point, after the turning point positive correlation has between air pollution and GDP. Moreover, a negative and significant relationship has been found between recycling and air pollution. According to the regression result, a 1% rise in recycling rates mitigates the (2.5 μm particulate matter) PM2.5 in between 0.021 and 0.024%.

Babel and Vilaysouk [17] estimated the relationship between greenhouse gas emission and municipal solid waste management in Vientiane by using IPCC and ABC EI model approaches. According to simulation results, each scenario reduces greenhouse gas emission at a different level, especially scenario 3 (recycling amount increased and recycled material market considered) where both reduce greenhouse gas emission and cause less air pollution. These results imply that municipal solid waste management help mitigates environmental hazards.

Vaverková et al. [156] discussed the nexus between municipal solid waste and the environment in Czech Republic Štěpánovice MSW landfill. The landfill leachate has been analyzed by phytotoxity tests and the results show there is no environmental hazard in the landfill area.

Magazzino et al. [99] explored the nexus between solid waste management, greenhouse gas emission, and GDP in Switzerland by using 1970 to 2017 data. Firstly, to find the causality between parameters, they used Toda-Yamamoto, Dolado-Lutkepohl, and Granger causality methods. Secondly, they utilized the machine learning technique (D2C algorithm) to investigate the relationship between variables. According to the analysis result, they found solid waste recycling and composting as the main factor to reduce greenhouse gas emissions. In addition to this there exists a bidirectional causality relationship between municipal solid waste and GDP. That shows that the EKC hypothesis is valid for Switzerland.

Boubellouta and Kusch-Brandt [27] examined the EKC hypothesis validity of e-waste in the year 2000–2016 for 30 European countries. They utilized three different methods—the generalized method of moments (GMM), the two-stage least square (2SLS) regression method, and the cross-section method. All methods' results showed that e-waste EKC is valid for European countries. That means that GDP growth results in more e-waste collection until the turning point. But after the turning point more GDP growth results in less e-waste collection.

Tominac et al. [150] investigated the environmental and economic usefulness of landfill strategies for municipal organic waste management in Milwaukee. They found that landfill infrastructure reduces greenhouse gas emissions' additional tax policies that positively affect the reduction level.

Lykov and Obolenskaya [97] investigated the effect of municipal solid waste (MSW) decomposition on the environment and economic factors from 2010 to 2020 in the Kaluga region. According to the result, there is clear evidence that the larger MSW landfills are the more feasible investment for economic conditions and more efficient for environmental protection.

Cerqueira et al. [33] researched the nexus between recycling rates, renewable energy, and economic development. They have been utilized in 28 OECD countries from 2000 to 2016 data. The three-stage least squares (3SLS) method have been used as an analyzing method. They found that recycling and renewable energy consumption are important parameters for economic and sustainable development issues. In addition to this, rising in recycling and renewable energy consumption results from fewer climate hazards and these results support the idea of more support of green policies and more economic development.

Razzaq et al. [129] examined the nexus between solid waste recycling, economic growth, and environmental quality using data from the USA for the period 1990–2017. They used bootstrap auto-regressive distributive lag (ARDL) and Granger causality methods. They found that a 1% increase in solid waste recycling results in a 0.317% increase in the long run and a 0.157% increase in a short period of economic growth. In addition to this, they found that a 1% increase in solid waste recycling mitigated 0.209% in a long period and 0.087% in a short period of carbon emissions. This result implies a unidirectional causality from municipal solid waste recycling to economic growth, carbon emissions, and energy efficiency.

Chakraborty et al. [34] investigated the relationship between waste management and income over the period 2001–2018 in 103 Italian provinces. They performed a panel cointegration test, dynamic panel threshold test, and panel GMM approach. The results showed that after the threshold point waste management is really important for the socio-economic condition. Moreover, it has been determined that waste management is better implemented in cities with strong economic activities and several tourists.

Magazzino and Falcone [98] observed the relationship between municipal waste generation, economic growth, and greenhouse gas emission by utilizing Switzerland data between 1990 and 2017. They employed dynamic ARDL and fuzzy cognitive map analysis methods. According to the analysis results, unidirectional causality was determined by municipal waste and economic growth to greenhouse gas emissions. And also, bidirectional causality has been determined between municipal waste and greenhouse gas emissions.

Similar to the previous papers, Boubellouta and Kusch-Brandt [26] analyzed the 30 European countries' data in the years 2008–2018. On the other hand, different other studies investigated the key factors of e-waste recycling. To test the EKC hypothesis, they used the stochastic impacts by regression on population, affluence, and technology (STIRPAT) method and also the panel quantile regression method to test the relations between parameters. According to the empirical results, N shape relationship has been found between GDP and e-waste recycling. That means, GDP and e-waste recycling positively correlated first then after the turning point they have a negative correlation between them and then after the last turning point again positive correlation has been found between parameters. Additionally, e-waste recycling positively affects economic development.

Kocak and Baglitas [87] examined the effect of economic, social, and technical parameters on MSW via the period of 2003–2018 OECD data. They have benefited from difference-GMM, system-GMM and the panel fixed effect model. They found the inverted-U-shaped EKC hypothesis as valid. Moreover, solid waste has a negative relationship with human development, financial development, and energy efficiency. In contrast with that result, credit expansion, income inequality, and poverty reduction are the increasing factors of solid waste.

Numerous studies have investigated the EKC hypothesis and revealed mixed results. For instance, [21, 37, 41, 79, 111, 137] confirmed the EKC hypothesis, while [9, 32, 46, 118, 132] disconfirmed the existence of EKC hypothesis. Some other studies have reached different conclusions regarding the shape of the relationship between environmental pollution and income. For example, [151] for Ghana, [35] for MENA countries, and [128] for Asian countries, concluded a U-shaped relationship between pollution and income. Akpan and Chuku [6] for Nigeria, [96] for OECD, and Danish and Wang [42] for BRICS countries, [13] for G20 economies found an N-shaped relationship between pollution and income. [20] for Malaysia, [8] for the Kingdom of Saudi Arabia revealed an inverted N-shaped shaped relationship between pollution and income.

Another important factor in environmental pollution is energy consumption. [24] for 17 OECD countries, [159] for Pakistan, [142] for 74 countries, [59] for seven countries, [164] for BRIICTS countries, [45] for G7 countries, [172] for China, [67] for developing countries, [167] for Argentina, H. [83] for BRICS, and [1] for Sweden supported that renewable energy consumption reduces environmental pollution. On the other hand, [77] for 69 countries, [43] for Pakistan, [14, 171] for Africa, [30] for African countries, [37] for N-11 economies concluded that non-renewable energy consumption contributes to pollution.

Besides, several other factors like human capital [92, 149, 165], urbanization [11, 101, 133], population [107, 139, 166], trade openness [44, 51, 71, 100, 168], globalization [60, 84, 91, 93, 167] and foreign direct investments [69, 75, 127] affect environmental degradation.

3 Empirical Strategy

3.1 Empirical Model and Data

This study investigates how waste management affects environmental pollution by controlling economic growth, renewable energy use, and trade openness. The countries and periods used for this analysis were chosen based on the availability of the dataset on the waste amount, recycling rate, and landfill rate. For this reason, (Austria, Belgium, Denmark, France, Germany, Hungary, Italy, Japan, Luxembourg, Netherlands, Norway, Portugal, Slovak, Spain, Sweden, United Kingdom, United States) were selected from OECD countries. Although the number of countries examined is less than the total number of OECD countries, we think that the result of this study can represent other OECD countries due to similar characteristics of OECD countries. This study uses the panel data of OECD countries from 1990 to 2016 and the following basic model:

$$EF_{it} = f\left(Y_{it}, \ Y_{it}^2, \ REC_{it}, \ TO_{it}, \ Waste_{it}, \ Recycling_{it}, \ Landfill_{it}\right) \quad (1)$$

In Eq. (1), EF is ecological footprint (as an indicator of environmental degradation); Y (as indicators of economic growth) is the real gross domestic product, Y^2 is the real gross domestic product per capita square; REC is the renewable energy consumption; TO is Trade openness; $Waste$ is Municipal waste amount; $Recycling$ is recycling rate, $Landfill$ is landfill rate.

The present study used a log-linear quadratic model. The logarithmic form of all variables is used in the empirical analysis.

$$\ln EF_{it} = \beta_0 + \beta_{1i} \ln Y_{it} + \beta_{2i} \ln Y_{it}^2 + \beta_{3i} \ln REC + \beta_{4i} \ln TO_{it}$$
$$+ \beta_{5i} \ln Waste_{it} + \beta_{6i} \ln Recycling_{it} + \beta_{7i} \ln Landfill_{it} + \varepsilon_{it} \quad (2)$$

Table 1 Summary of the variables

Variables	Unit	Source
Ecological footprint ($\ln EF$)	Global hectares per person	Global Footprint Network (2022)
Economic growth ($\ln Y$)	GDP per capita in constant 2015 US$	World Development Indicators, World Bank (2022)
Renewable energy consumption ($\ln REC$)	Share of renewable energy in total final energy consumption	World Development Indicators, World Bank (2022)
Trade openness ($\ln TO$)	Sum of exports and imports of goods and services measured as a share of gross domestic product	World Development Indicators, World Bank (2022)
Amount of waste ($\ln Waste$)	Per capita (Kilogram)	OECD Environment Statistics (2022)
Recycling rate ($\ln Recycling$)	Percentage of total waste amount	OECD Environment Statistics (2022)
Landfill rate ($\ln Landfill$)	Percentage of total waste amount	OECD Environment Statistics (2022)

In Eq. (2) in line with the principles of the EKC hypothesis, the signs of the elasticity parameters to β_1 and β_2 are expected to be positive and negative, respectively [18, 23, 25, 38, 48, 64, 65, 145]. Because renewable energy consumption mitigates environmental degradation, β_3 is expected to be negative [11, 36, 40, 85, 144]. The parameter β_4 may be either positive or negative [31, 134, 140, 161]. Moreover, the signs of the elasticity parameter β_5 and β_7 are likely to be positive as well since boosting environmental degradation [27, 76, 112]. The sign of the elasticity parameter β_6 is likely to be negative as well since curb environmental degradation [26, 33, 55].

The explanations of the variables and the source from which the data set is obtained are given in Table 1.

Descriptive statistics for each of the variables are reported in Table 2. The square of economic growth has the highest mean value of 109.761 with a minimum of 79.706 and a maximum of 135.247 over the sample period. In terms of mean average, renewable energy, recycling rate, and landfill rate are close to each other at 2.249, 2.862, and 2.783 respectively. The average ecological footprint, economic growth, trade openness, and amount of waste are 1.777, 10.461, 4.349, and 6.243 respectively. Jarque–Bera test statistics indicated that the variables do not follow the normal distribution.

Table 3 presents the correlation matrix for the analyzed variables. While economic growth, trade openness, waste and recycling are positively correlated, renewable energy and landfill are negatively correlated.

Table 2 Summary of descriptive statistics

	ln EF	ln Y	ln Y^2	ln REC	ln TO	ln $Waste$	ln $Recycling$	ln $Landfill$
Mean	1.777	10.461	109.761	2.249	4.349	6.243	2.862	2.783
Median	1.732	10.536	111.017	2.208	4.293	6.247	3.106	3.325
Maximum	2.875	11.629	135.247	4.112	5.886	6.757	3.894	4.507
Minimum	1.077	8.927	79.706	−0.159	2.796	5.472	−0.059	−1.604
Std. Dev	0.327	0.558	11.495	1.020	0.612	0.245	0.773	1.598
Skewness	1.027	−0.637	−0.462	−0.115	0.021	−0.395	−1.841	−0.890
Kurtosis	4.360	3.538	3.411	2.516	3.057	3.314	6.270	2.642
Jarque–Bera	103.227	32.5248	17.411	4.879	0.087	12.299	412.556	56.088
Probability	0.000	0.000	0.000	0.087	0.957	0.002	0.000	0.000
Observations	408	408	408	408	408	408	408	408

Table 3 Correlation matrix

	ln EF	ln Y	ln Y^2	ln REC	ln TO	ln $Waste$	ln $Recycling$	ln $Landfill$
ln EF	1							
ln$Y$	0.785	1						
ln Y^2	0.795	0.999	1					
ln REC	−0.183	0.093	0.093	1				
ln TO	0.224	0.032	0.053	0.004	1			
ln $Waste$	0.604	0.632	0.630	−0.027	−0.076	1		
ln $Recycling$	0.403	0.709	0.695	0.179	0.019	0.495	1	
ln $Landfill$	−0.184	−0.447	−0.441	−0.182	−0.144	−0.134	−0.571	1

3.2 Methodology

3.2.1 Cross-Sectional Dependence and Heterogeneity Test

Ignoring the cross-section dependency will result in the selection of an inappropriate methodological approach. This might render other following analysis' results and findings erroneous [110, 113]. Cross-sectional dependence tests are important because they are enabled to choose the appropriate empirical model to conduct the unit root, and cointegration estimation [54, 174]. Therefore, this study utilizes Lagrange Multiplier (LM) developed by [29], CD and CD_{LM} tests were suggested by [121], and LM_{adj} developed by [123] tests to determine the presence or absence of cross-sectional dependence. The null hypothesis of cross-section dependence tests indicates that there is no cross-section dependence. $\tilde{\Delta}$ and $\tilde{\Delta}_{adj}$ tests developed by

[123] are used to investigate the presence of slope heterogeneity. Heterogeneity tests search for the null hypothesis of slope homogeneity.

3.2.2 Panel Unit Root Test

In the presence of a unit root in the series, a spurious regression problem will arise in the panel data set [81, 141, 146]. Therefore, it is important to check the unit root before estimating cointegration and regression techniques [170]. The study applies the cross-sectional IPS (CIPS) second-generation panel unit root test developed by Pesaran. This test gives efficient stationarity results since it takes into account cross-section dependence and panel heterogeneity [19, 49]. The null hypothesis of the CIPS test is that there is a unit root.

3.2.3 Panel Cointegration Test

The study used the [162] Durbin-Hausman cointegration test to investigate the long-run cointegration between the variables. One of the advantages of the Durbin-Hausman panel cointegration test is that it does not require any prior information about the integration level of the variables [10]. This test also considers cross-sectional dependence and heterogeneity [10, 16]. The test is based on two statistics, a group statistic (DH_g) and a panel statistic (DH_p). Durbin-Hausman test allows autoregressive parameters to be homogeneous (DH_p) or heterogeneous (DH_g). The null hypothesis of the Durbin-Hausman test is that there is no cointegration among the variables [4, 73, 154].

3.2.4 Panel Coefficient Estimator

The long-term effects of the independent variables on the dependent variable are obtained with the fully modified ordinary least-square (FMOLS) recommended by [119] and pooled mean group (PMG) suggested by [122]. Our model consists of 17 countries and 27 years. Since our sample is small, FMOLS and PMG estimators are applied, which give consistent results in small samples. FMOLS technique is non-parametric and it controls the autocorrelation problem [116, 120]. The FMOLS estimator is suitable for studies with a small sample size [24]. The PMG model can be applied while the dependent variable is I(1) regardless of whether the independent variables are I(1) or I(0) [102, 109]. PMG model can solve the endogeneity, heteroscedasticity, autocorrelation, and multicollinearity problems econometrically [160]. PMG model generates consistent estimates in small samples [2].

4 Empirical Findings

In Table 4, the cross-sectional dependence results are depicted. The results of the cross-section test revealed the existence of cross-section dependence in all variables. The presence of cross-section dependence in the data means that the OECD countries share common characteristics and shocks. So, shocks are transmitted across OECD countries. Also, the results of heterogeneity tests are presented in Table 4. The analysis of heterogeneity across the cross-sections validated the presence of heterogeneity across the samples. This implies that the OECD countries exhibit unique economic characteristics.

The study utilizes second-generation unit root tests to be applied since cross-section dependency is determined. CIPS unit root test, which allows cross-section dependence, was used to observe the stationary properties of the series. The results of the CIPS unit root test are illustrated in Table 5. The outcomes of the CIPS test indicate that all variables are not stationary at the level. However, it has been observed that all variables become stationary in the first difference.

The null hypothesis for the [162] cointegration test indicated that the lack of cointegration between variables was rejected. This finding means that there is a long-run

Table 4 Results of cross-sectional dependence test and homogeneity tests

Variables	LM		CD_{LM}		CD		LM_{adj}	
	C	C + T	C	C + T	C	C + T	C	C + T
ln EF	211.767[a] (0.000)	222.864[a] (0.000)	4.594[a] (0.000)	5.267[a] (0.000)	−1.270 (0.102)	−1.621[c] (0.053)	1.056 (0.146)	0.156 (0.438)
ln Y	230.685[a] (0.000)	222.769[a] (0.000)	5.741[a] (0.000)	5.261[a] (0.000)	2.481[a] (0.007)	1.353[c] (0.088)	4.912[a] (0.000)	4.270[a] (0.000)
ln Y^2	230.162[a] (0.000)	223.351[a] (0.000)	5.709[a] (0.000)	5.296[a] (0.000)	−2.443[a] (0.007)	−1.379[c] (0.084)	5.083[a] (0.000)	4.393[a] (0.000)
ln REC	221.323[a] (0.000)	248.952[a] (0.000)	5.173[a] (0.000)	6.849[a] (0.000)	−2.279[b] (0.011)	−1.353[c] (0.088)	−0.152 (0.561)	0.049 (0.481)
ln TO	201.941[a] (0.000)	210.415[a] (0.000)	3.998[a] (0.000)	4.512[a] (0.000)	−2.537[a] (0.006)	−2.413[c] (0.008)	0.763 (0.223)	0.697 (0.243)
ln $Waste$	198.299[a] (0.000)	217.286[a] (0.000)	3.777[a] (0.000)	4.929[a] (0.000)	−1.330[c] (0.092)	−1.609[c] (0.054)	−2.445 (0.993)	−2.835 (0.998)
ln $Recycling$	202.557[a] (0.000)	207.719[a] (0.000)	4.036[a] (0.000)	4.349[a] (0.000)	−1.255 (0.105)	−1.114 (0.133)	−1.438 (0.925)	−0.891 (0.814)
ln $Landfill$	158.340[c] (0.092)	177.787[a] (0.009)	1.355[c] (0.088)	2.534[a] (0.006)	−1.177 (0.120)	−1.019 (0.154)	0.651 (0.258)	1.363[c] (0.086)

Homogeneity Tests								
Test	Statistic	Prob						
$\tilde{\Delta}$	7.631	0.000						
$\tilde{\Delta}_{adj}$	9.346	0.000						

Note [a],[b] and [c] indicate statistical significance at 10, 5, and 1% levels, respectively. C, C + T indicate Constant, Constant, and Trend, respectively

Table 5 Results of panel unit root tests

Variables	Level		First difference	
	C	C + T	C	C + T
ln EF	−2.097	−2.414	−4.806[a]	−4.806[a]
ln Y	−2.112	−2.396	−3.373[a]	−3.392[a]
ln Y^2	−2.072	−2.373	−3.277[a]	−3.389[a]
ln REC	−1.881	−2.828	−5.102[a]	−5.200[a]
ln TO	−1.806	−2.038	−3.507[a]	−3.818[a]
ln $Waste$	−1.312	−2.450	−4.093[a]	−4.525[a]
ln Re $cycling$	−2.516[a]	−2.595	−3.806[a]	−3.806[a]
ln $Landfill$	−1.337	−2.007	−3.659[a]	−4.469[a]

Note Critical values for 1, 5, and 10% levels are − 2.380, − 2.200, and − 2.110, respectively in the constant model. Critical values for 1, 5, and 10% levels are −2.880, −2.720, and -2.630 in the constant and trend model. a indicates statistical significance at the 1% level

Table 6 Results of panel cointegration test

	Constant	Constant and Trend
DH_g	295.278[a] (0.000)	162.979[a] (0.000)
DH_p	2.960[a] (0.002)	1.884[a] (0.030)

Note [a]indicates statistical significance at 1% level

relationship between ecological footprint, growth, renewable energy consumption, trade openness, waste, recycling, and landfill, and thus, the long-run impacts of explanatory variables on ecological footprint should be investigated.

Table 7, presents the FMOLS, and PMG outcomes. The economic growth impact on the ecological footprint is positive and significant. Therefore, a 1% raise in economic growth raises ecological footprint by 4.001% in this instance of the FMOLS estimator, and 4.239% in this instance of the PMG estimator. Furthermore, the economic growth square impact on ecological footprint is negative and significant. Therefore, there is a 1% raise in economic growth square and decrease in ecological footprint by 0.176% in the case of the FMOLS estimator, and 0.167% in the case of the PMG estimator. This result means that there is an inverted U-shaped connection between economic growth and environmental degradation. This outcome corroborates the findings of [24, 39, 61, 78, 90] for OECD, [131] for 170 countries, [130] for 103 countries, [21] for E7 countries, [104] for South Asia, [157] for Southeastern Europe, [135] for European countries. Nonetheless, the outcome contradicts the studies of [47, 58, 68, 158] for OECD countries, [175] for 5 ASEAN countries, [94] for ASEAN countries, [15] for 31 countries, [72] for emerging economies, [50] for BRICST, [66] for E7 countries, who found the invalid EKC hypothesis.

Table 7 Results of panel long-run coefficient estimator

	FMOLS			PMG		
	Coefficient	t-stat	p-value	Coefficient	z-stat	p-value
ln Y	4.001[a]	5.451	0.000	4.239[a]	3.94	0.000
ln Y^2	−0.176[a]	−4.794	0.000	−0.167[a]	−3.01	0.003
ln REC	−0.117[a]	−7.591	0.000	−0.211[a]	−9.90	0.000
ln TO	−0.194[a]	−3.403	0.000	−0.378[a]	−8.69	0.000
ln $Waste$	0.165[a]	3.141	0.001	0.164[a]	3.09	0.002
ln $Recycling$	−0.049[a]	−3.273	0.001	−0.026[a]	−2.88	0.004
ln $Landfill$	−0.015[c]	−1.938	0.053	0.014	1.62	0.106

Note [a], [b] and [c] indicate statistical significance at 10, 5, and 1% levels, respectively

The effect of renewable energy consumption on ecological footprint is negative and significant. In this case, a 1% increase in renewable energy consumption decreased ecological footprint by 0.117% in the case of the FMOLS estimator, and 0.211% in the case of the PMG estimator. This result means that renewable energy consumption decreases environmental degradation. This outcome is consistent with those of [136] for South Africa, [74] for sub-Saharan Africa, [169] for OECD, [138] for OECD, [40] for OECD, [117] for BRIC countries, [5] for G20 nations, [124] for European countries, [126] for G20 countries, [108] and for Africa, who established an inverse relationship between renewable energy consumption and environmental degradation.

Moreover, the effect of trade openness on ecological footprint is significant and negative, suggesting that a 1% raise in trade openness reduces ecological footprint by 0.194% in the case of the FMOLS estimator, and 0.378% in the case of the PMG estimator. This is consistent with the results of the studies [134] for OECD, [63] OECD, [86] for OECD, [115] for the European Union, [31] for ASEAN economies, [7] for Europe, [95] for Pakistan, [114] for 144 countries, [173] for ten newly industrialized countries. Nonetheless, the outcome contradicts the studies of [140] for global, high, middle, and low-income countries, [147] for European Union, [155] for 66 developing countries, [161] for Asian economies, [3] for OECD, [42] for BRICS, who found a positive relationship between trade openness and environmental degradation.

Furthermore, when the impact of waste management on the ecological footprint is evaluated because the ecological footprint reflects the demand for productive areas to generate resources and absorb carbon dioxide emissions, recycling can reduce the ecological footprint by reducing the area required to absorb carbon dioxide emissions. If municipal waste and landfills previously occupy a biologically productive area, these materials can damage ecosystems when released into the environment [62].

The effect of waste on ecological footprint is positive and significant, which illustrates that an upsurge in waste contributes to environmental degradation. Therefore, a 1% increase in waste triggers an ecological footprint of 0.165% in the case of the

FMOLS estimator and 0.164% in the case of the PMG estimator. This outcome is consistent with the studies of [98] for Switzerland.

Moreover, the recycling-ecological footprint association is negative and significant. Therefore, a 1% increase in recycling decreases ecological footprint by 0.049% in the case of the FMOLS estimator, and 0.026% in the case of the PMG estimator. This empirical finding is supported by [33] for OECD countries, [55] for Massachusetts, [17] for Lao PDR, [99] for Switzerland, [129] for the United States.

Lastly, the effect of landfill on ecological footprint is significant and negative in the case of the FMOLS estimator and insignificant in the case of the PMG estimator [76] and [112] stated that landfills have negative effects on the environment.

5 Conclusions and Policy Implications

The current study is presented to contribute to the literature by investigating the long-term effects of municipal waste, recycling, and landfills on the ecological footprint in 17 OECD countries from 1995–2018. The dependent variable in this study is ecological footprint while the independent variables are growth, renewable energy consumption, municipal waste, recycling, and landfills. To achieve a comprehensive finding, this study obtains comparative results by using different estimation methods. In addition, this study incorporates two control variables, namely, globalization and economic growth. The study employs [162] cointegration test to detect the long-run relationship between ecological footprint and regressors. Finally, the study also employs FMOLS and PMG estimators to estimate the long-term impact of explanatory variables on the ecological footprint.

The results of the empirical study for OECD countries show that the EKC hypothesis is valid. Furthermore, renewable energy consumption, trade openness, and recycling reduce the ecological footprint in these countries. Mixed results have been obtained for landfill. Landfills affect ecological footprint negatively in the case of the fully modified ordinary least-square (FMOLS) estimator. This is because the negative environmental impact of landfills in OECD countries is reduced by applying appropriate technology and effective solid waste management. Landfills affect ecological footprint insignificant in the case of pooled mean group (PMG) estimator.

As long as humanity exists, production and consumption will continue. Therefore, waste will arise as a result of production and consumption. Some of these wastes can be recycled and reused. The other part is sent to landfills. Landfills are an important source of pollution as they produce methane gas and carbon dioxide emissions, and the garbage there will be a problem for future generations as it decomposes at a slow rate. Waste management must be managed effectively so that a larger portion of the waste is recycled and a smaller portion is sent to landfills.

This study's scope was limited to OECD countries and only a few waste management variables were considered when investigating the effect of waste, recycling, and landfills on ecological footprint. For future studies, recovery, material recovery, and incineration can be incorporated into the environmental degradation model.

Additionally, evaluation of the sample of global economies, different regions, and various income-wise groups of economies will be constructive for more specific policy implications.

References

1. Adebayo TS, Rjoub H, Akinsola GD, Oladipupo SD (2022) The asymmetric effects of renewable energy consumption and trade openness on carbon emissions in Sweden: new evidence from quantile-on-quantile regression approach. Environ Sci Pollut Res 29:1875–1886. https://doi.org/10.1007/S11356-021-15706-4/FIGURES/4
2. Adedoyin FF, Alola AA, Bekun FV (2020) The nexus of environmental sustainability and agro-economic performance of Sub-Saharan African countries. Heliyon 6:e04878. https://doi.org/10.1016/J.HELIYON.2020.E04878
3. Ahmad M, Khan Z, Rahman ZU et al (2019) Can innovation shocks determine CO_2 emissions (CO_2e) in the OECD economies? A new perspective. Econ Innov New Technol 30:89–109. https://doi.org/10.1080/10438599.2019.1684643
4. Ahmed Z, Adebayo TS, Udemba EN et al (2022) Effects of economic complexity, economic growth, and renewable energy technology budgets on ecological footprint: the role of democratic accountability. Environ Sci Pollut Res 29:24925–24940. https://doi.org/10.1007/S11356-021-17673-2
5. Ajide KB, Mesagan EP (2022) Heterogeneous analysis of pollution abatement via renewable and non-renewable energy: lessons from investment in G20 nations. Environ Sci Pollut Res 29:36533–36546. https://doi.org/10.1007/S11356-022-18771-5/TABLES/9
6. Akpan UF, Chuku A (2011) Economic growth and environmental degradation in Nigeria: beyond the environmental kuznets curve. In: Forthcoming in procedings of 2011 annual conference of NAEE. Munich Personal RePEc Archive, Abuja
7. Al-Mulali U, Ozturk I, Lean HH (2015) The influence of economic growth, urbanization, trade openness, financial development, and renewable energy on pollution in Europe. Nat Hazards 79:621–644. https://doi.org/10.1007/S11069-015-1865-9
8. Aljadani A, Toumi H, Toumi S et al (2021) Investigation of the N-shaped environmental Kuznets curve for COVID-19 mitigation in the KSA. Environ Sci Pollut Res 28:29681–29700. https://doi.org/10.1007/S11356-021-12713-3/TABLES/7
9. Alshehry AS, Belloumi M (2017) Study of the environmental Kuznets curve for transport carbon dioxide emissions in Saudi Arabia. Renew Sustain Energy Rev 75:1339–1347. https://doi.org/10.1016/J.RSER.2016.11.122
10. Amin S, Jamasb T, Nepal R (2021) Regulatory reform and the relative efficacy of government versus private investment on energy consumption in South Asia. Econ Anal Policy 69:421–433. https://doi.org/10.1016/J.EAP.2020.12.019
11. Anwar A, Sinha A, Sharif A et al (2022) The nexus between urbanization, renewable energy consumption, financial development, and CO2 emissions: evidence from selected Asian countries. Environ Dev Sustain 24:6556–6576. https://doi.org/10.1007/S10668-021-01716-2/TABLES/6
12. Ari I, Şentürk H (2020) The relationship between GDP and methane emissions from solid waste: a panel data analysis for the G7. Sustain Prod Consum 23:282–290. https://doi.org/10.1016/J.SPC.2020.06.004
13. Awan AM, Azam M (2021) Evaluating the impact of GDP per capita on environmental degradation for G-20 economies: does N-shaped environmental Kuznets curve exist? Environ Dev Sustain 1–24. https://doi.org/10.1007/S10668-021-01899-8/FIGURES/6
14. Awodumi OB, Adewuyi AO (2020) The role of non-renewable energy consumption in economic growth and carbon emission: evidence from oil producing economies in Africa. Energy Strateg Rev 27:100434. https://doi.org/10.1016/J.ESR.2019.100434

15. Aye GC, Edoja PE (2017) Effect of economic growth on CO_2 emission in developing countries: Evidence from a dynamic panel threshold model 5. http://www.editorialmanager.com/cogent econ. https://doi.org/10.1080/23322039.2017.1379239

16. Azam M, Awan AM (2022) Health is wealth: a dynamic SUR approach of examining a link between climate changes and human health expenditures. Soc Indic Res 1–24. https://doi.org/ 10.1007/S11205-022-02904-X/TABLES/7

17. Babel S, Vilaysouk X (2016) Greenhouse gas emissions from municipal solid waste management in Vientiane, Lao PDR. Waste Manag Res 34:30–37. https://doi.org/10.1177/073424 2X15615425

18. Balsalobre-Lorente D, Ibáñez-Luzón L, Usman M, Shahbaz M (2022) The environmental Kuznets curve, based on the economic complexity, and the pollution haven hypothesis in PIIGS countries. Renew Energy 185:1441–1455. https://doi.org/10.1016/J.RENENE.2021. 10.059

19. Bano S, Liu L, Khan A (2022) Dynamic influence of aging, industrial innovations, and ICT on tourism development and renewable energy consumption in BRICS economies. Renew Energy 192:431–442. https://doi.org/10.1016/J.RENENE.2022.04.134

20. Bekhet HA, Othman NS (2018) The role of renewable energy to validate dynamic interaction between CO_2 emissions and GDP toward sustainable development in Malaysia. Energy Econ 72:47–61. https://doi.org/10.1016/J.ENECO.2018.03.028

21. Bekun FV, Gyamfi BA, Onifade ST, Agboola MO (2021) Beyond the environmental Kuznets Curve in E7 economies: accounting for the combined impacts of institutional quality and renewables. J Clean Prod 314: https://doi.org/10.1016/J.JCLEPRO.2021.127924

22. Bethel A, Ethan B, Jordan H et al (2021) Pollution vs waste - energy education. In: Energy Educ. https://energyeducation.ca/encyclopedia/Pollution_vs_waste. Accessed 28 July 2022

23. Bilgili F, Khan M, Awan A (2022) Is there a gender dimension of the environmental Kuznets curve? Evidence from Asian countries. Environ Dev Sustain 2022:1–32. https://doi.org/10. 1007/S10668-022-02139-3

24. Bilgili F, Koçak E, Bulut Ü (2016) The dynamic impact of renewable energy consumption on CO2 emissions: a revisited Environmental Kuznets Curve approach. Renew Sustain Energy Rev 54:838–845. https://doi.org/10.1016/J.RSER.2015.10.080

25. Bilgili F, Nathaniel SP, Kuşkaya S, Kassouri Y (2021) Environmental pollution and energy research and development: an Environmental Kuznets Curve model through quantile simulation approach. Environ Sci Pollut Res 28:53712–53727. https://doi.org/10.1007/S11356-021-14506-0/TABLES/2

26. Boubellouta B, Kusch-Brandt S (2022) Driving factors of e-waste recycling rate in 30 European countries: new evidence using a panel quantile regression of the EKC hypothesis coupled with the STIRPAT model. Environ Dev Sustain 1–28. https://doi.org/10.1007/S10668-022-02356-W/FIGURES/1

27. Boubellouta B, Kusch-Brandt S (2020) Testing the environmental Kuznets Curve hypothesis for E-waste in the EU28+2 countries. J Clean Prod 277:123371. https://doi.org/10.1016/J. JCLEPRO.2020.123371

28. Boulder County (2022) Reduce, Reuse & recycle - boulder county. https://bouldercounty.gov/ environment/recycle/reduce-reuse-recycle/. Accessed 26 July 2022

29. Breusch TS, Pagan AR (1980) The Lagrange multiplier test and its applications to model specification in econometrics. Rev Econ Stud 47:239. https://doi.org/10.2307/2297111

30. Brini R (2021) Renewable and non-renewable electricity consumption, economic growth and climate change: evidence from a panel of selected African countries. Energy 223:120064. https://doi.org/10.1016/J.ENERGY.2021.120064

31. Burki U, Tahir M (2022) Determinants of environmental degradation: evidenced-based insights from ASEAN economies. J Environ Manage 306:114506. https://doi.org/10.1016/ J.JENVMAN.2022.114506

32. Caglar AE, Mert M, Boluk G (2021) Testing the role of information and communication technologies and renewable energy consumption in ecological footprint quality: evidence from world top 10 pollutant footprint countries. J Clean Prod 298:126784. https://doi.org/10. 1016/J.JCLEPRO.2021.126784

33. Cerqueira PA, Soukiazis E, Proença S (2021) Assessing the linkages between recycling, renewable energy and sustainable development: evidence from the OECD countries 23:9766–9791. https://doi.org/10.1007/s10668-020-00780-4
34. Chakraborty SK, Mazzanti M, Mazzarano M (2022) Municipal Solid Waste generation dynamics. Breaks and thresholds analysis in the Italian context. Waste Manag 144:468–478. https://doi.org/10.1016/J.WASMAN.2022.04.022
35. Charfeddine L, Mrabet Z (2017) The impact of economic development and social-political factors on ecological footprint: a panel data analysis for 15 MENA countries. Renew Sustain Energy Rev 76:138–154. https://doi.org/10.1016/J.RSER.2017.03.031
36. Cheng C, Ren X, Wang Z, Yan C (2019) Heterogeneous impacts of renewable energy and environmental patents on CO_2 emission - evidence from the BRIICS. Sci Total Environ 668:1328–1338. https://doi.org/10.1016/J.SCITOTENV.2019.02.063
37. Chien F (2022) How renewable energy and non-renewable energy affect environmental excellence in N-11 economies? Renew Energy 196:526–534. https://doi.org/10.1016/J.RENENE.2022.07.013
38. Churchill SA, Inekwe J, Ivanovski K, Smyth R (2018) The environmental Kuznets curve in the OECD: 1870–2014. Energy Econ 75:389–399. https://doi.org/10.1016/J.ENECO.2018.09.004
39. Cole MA (2004) Trade, the pollution haven hypothesis and the environmental Kuznets curve: examining the linkages. Ecol Econ 48:71–81. https://doi.org/10.1016/J.ECOLECON.2003.09.007
40. Dagar V, Khan MK, Alvarado R et al (2022) Impact of renewable energy consumption, financial development and natural resources on environmental degradation in OECD countries with dynamic panel data. Environ Sci Pollut Res 29:18202–18212. https://doi.org/10.1007/S11356-021-16861-4/TABLES/6
41. Danish UR (2022) Analyzing energy innovation-emissions nexus in China: a novel dynamic simulation method. Energy 244:123010. https://doi.org/10.1016/J.ENERGY.2021.123010
42. Danish WZ (2019) Does biomass energy consumption help to control environmental pollution? Evidence from BRICS countries. Sci Total Environ 670:1075–1083. https://doi.org/10.1016/J.SCITOTENV.2019.03.268
43. Danish ZB, Wang B, Wang Z (2017) Role of renewable energy and non-renewable energy consumption on EKC: evidence from Pakistan. J Clean Prod 156:855–864. https://doi.org/10.1016/J.JCLEPRO.2017.03.203
44. Dauda L, Long X, Mensah CN et al (2021) Innovation, trade openness and CO2 emissions in selected countries in Africa. J Clean Prod 281:125143. https://doi.org/10.1016/J.JCLEPRO.2020.125143
45. Destek MA, Aslan A (2020) Disaggregated renewable energy consumption and environmental pollution nexus in G-7 countries. Renew Energy 151:1298–1306. https://doi.org/10.1016/J.RENENE.2019.11.138
46. Destek MA, Sinha A (2020) Renewable, non-renewable energy consumption, economic growth, trade openness and ecological footprint: evidence from organisation for economic co-operation and development countries. J Clean Prod 242:118537. https://doi.org/10.1016/J.JCLEPRO.2019.118537
47. Dijkgraaf E, Vollebergh HRJ (2005) A test for parameter homogeneity in CO_2 panel EKC estimations. Environ Resour Econ 322(32):229–239. https://doi.org/10.1007/S10640-005-2776-0
48. Dinda S (2004) Environmental Kuznets curve hypothesis: a survey. Ecol Econ 49:431–455. https://doi.org/10.1016/J.ECOLECON.2004.02.011
49. Dogan E, Seker F (2016) An investigation on the determinants of carbon emissions for OECD countries: empirical evidence from panel models robust to heterogeneity and cross-sectional dependence. Environ Sci Pollut Res 23:14646–14655. https://doi.org/10.1007/S11356-016-6632-2/TABLES/7

50. Dogan E, Ulucak R, Kocak E, Isik C (2020) The use of ecological footprint in estimating the environmental Kuznets curve hypothesis for BRICST by considering cross-section dependence and heterogeneity. Sci Total Environ 723:138063. https://doi.org/10.1016/J.SCITOT ENV.2020.138063

51. Dou Y, Zhao J, Malik MN, Dong K (2021) Assessing the impact of trade openness on CO_2 emissions: evidence from China-Japan-ROK FTA countries. J Environ Manag 296:113241. https://doi.org/10.1016/J.JENVMAN.2021.113241

52. EEA (2014) Waste: a problem or a resource? In: EEA Signals. https://www.eea.europa.eu/publications/signals-2014/articles/waste-a-problem-or-a-resource. Accessed 28 July 2022

53. EEA (2013) Managing municipal solid waste : a review of achievements in 32 European countries. Publications Office

54. Ehigiamusoe KU, Dogan E (2022) The role of interaction effect between renewable energy consumption and real income in carbon emissions: evidence from low-income countries. Renew Sustain Energy Rev 154: https://doi.org/10.1016/J.RSER.2021.111883

55. Eleftherios G (2014) Relationship between recycling rate and air pollution: evidence from waste management municipality survey in the state of Massachusetts. SSRN Electron J. https://doi.org/10.2139/SSRN.2479296

56. EPA (2021) Recycling basics. In: United states environment. Prot. Agency. https://www.epa.gov/recycle/recycling-basics. Accessed 26 July 2022

57. EPA (2022) National recycling strategy. In: United states environment prot. Agency. https://www.epa.gov/recyclingstrategy. Accessed 28 July 2022

58. Erdogan S, Okumus I, Guzel AE (2020) Revisiting the environmental Kuznets curve hypothesis in OECD countries: the role of renewable, non-renewable energy, and oil prices. Environ Sci Pollut Res 27:23655–23663. https://doi.org/10.1007/S11356-020-08520-X/TABLES/5

59. Eyuboglu K, Uzar U (2020) Examining the roles of renewable energy consumption and agriculture on CO_2 emission in lucky-seven countries. Environ Sci Pollut Res 27:45031–45040. https://doi.org/10.1007/S11356-020-10374-2/TABLES/9

60. Farooq S, Ozturk I, Majeed MT, Akram R (2022) Globalization and CO_2 emissions in the presence of EKC: a global panel data analysis. Gondwana Res 106:367–378. https://doi.org/10.1016/J.GR.2022.02.002

61. Galeotti M, Lanza A (1999) Richer and cleaner? A study on carbon dioxide emissions in developing countries. Energy Policy 27:565–573. https://doi.org/10.1016/S0301-4215(99)000 47-6

62. Global Foodprint Network (2022) Open data platform. https://data.footprintnetwork.org/?_ga=2.184454117.2144727621.1652772150-2019296484.1555490724#/countryTrends?cn=5001&type=BCtot,EFCtot. Accessed 6 July 2022

63. Gozgor G (2017) Does trade matter for carbon emissions in OECD countries? Evidence from a new trade openness measure. Environ Sci Pollut Res 24:27813–27821. https://doi.org/10.1007/S11356-017-0361-Z/TABLES/5

64. Grossman GM, Krueger AB (1991) Environmental impacts of a North American free trade agreement. NBER Work Pap. https://doi.org/10.3386/W3914

65. Grossman GM, Krueger AB (1995) Economic growth and the environment. Q J Econ 110:353–377. https://doi.org/10.2307/2118443

66. Gyamfi BA, Adedoyin FF, Bein MA, Bekun FV (2021) Environmental implications of N-shaped environmental Kuznets curve for E7 countries. Environ Sci Pollut Res 28:33072–33082. https://doi.org/10.1007/S11356-021-12967-X/TABLES/9

67. Haldar A, Sethi N (2021) Effect of institutional quality and renewable energy consumption on CO_2 emissions−an empirical investigation for developing countries. Environ Sci Pollut Res 28:15485–15503. https://doi.org/10.1007/S11356-020-11532-2/TABLES/12

68. Halkos GE, Tzeremes NG (2009) Exploring the existence of Kuznets curve in countries' environmental efficiency using DEA window analysis. Ecol Econ 68:2168–2176. https://doi.org/10.1016/J.ECOLECON.2009.02.018

69. Haug AA, Ucal M (2019) The role of trade and FDI for CO2 emissions in Turkey: nonlinear relationships. Energy Econ 81:297–307. https://doi.org/10.1016/J.ENECO.2019.04.006

70. Hoornweg D, Bhada-Tata P (2012) Urban development series knowledge papers: what a waste : a global review of solid waste management, The World Bank. The World Bank, Washington, DC

71. Hossain SM (2011) Panel estimation for CO2 emissions, energy consumption, economic growth, trade openness and urbanization of newly industrialized countries. Energy Policy 39:6991–6999. https://doi.org/10.1016/J.ENPOL.2011.07.042

72. Hove S, Tursoy T (2019) An investigation of the environmental Kuznets curve in emerging economies. J Clean Prod 236: https://doi.org/10.1016/J.JCLEPRO.2019.117628

73. Ibrahim RL, Ajide KB (2021) Nonrenewable and renewable energy consumption, trade openness, and environmental quality in G-7 countries: the conditional role of technological progress. Environ Sci Pollut Res 28:45212–45229. https://doi.org/10.1007/S11356-021-13926-2/TABLES/11

74. Inglesi-Lotz R, Dogan E (2018) The role of renewable versus non-renewable energy to the level of CO_2 emissions a panel analysis of sub- Saharan Africa's Big 10 electricity generators. Renew Energy 123:36–43. https://doi.org/10.1016/J.RENENE.2018.02.041

75. Iqbal A, Tang X, Rasool SF (2022) Investigating the nexus between CO_2 emissions, renewable energy consumption, FDI, exports and economic growth: evidence from BRICS countries. Environ Dev Sustain 1–30. https://doi.org/10.1007/S10668-022-02128-6/TABLES/7

76. Iravanian A, Ravari SO (2020) Types of contamination in landfills and effects on the environment: a review study. In: IOP conference series: earth and environmental science. IOP Publishing, p 12083

77. Ben JM, Ben YS (2015) Output, renewable and non-renewable energy consumption and international trade: evidence from a panel of 69 countries. Renew Energy 83:799–808. https://doi.org/10.1016/J.RENENE.2015.04.061

78. Ben JM, Ben YS, Ozturk I (2016) Testing environmental Kuznets curve hypothesis: the role of renewable and non-renewable energy consumption and trade in OECD countries. Ecol Indic 60:824–831. https://doi.org/10.1016/J.ECOLIND.2015.08.031

79. Jena PK, Mujtaba A, Joshi DPP et al (2022) (2022) Exploring the nature of EKC hypothesis in Asia's top emitters: role of human capital, renewable and non-renewable energy consumption. Environ Sci Pollut Res 1:1–20. https://doi.org/10.1007/S11356-022-21551-W

80. Jeng SY, Lin CW, Tseng ML, Jantarakolica T (2020) Cradle-to-cradle zero discharge production planning system for the pulp and paper industry using a fuzzy hybrid optimization model. Manag Environ Qual An Int J 31:645–663. https://doi.org/10.1108/MEQ-06-2019-0120

81. Jun W, Mughal N, Zhao J, et al (2021) Does globalization matter for environmental degradation? Nexus among energy consumption, economic growth, and carbon dioxide emission. Energy Policy 153: https://doi.org/10.1016/J.ENPOL.2021.112230

82. Kaza S, Yao LC, Bhada-Tata P, Van Woerden F (2018) Urban development series: what a waste 2.0 a global snapshot of solid waste management to 2050. World Bank, Washington, DC

83. Khan H, Weili L, Khan I (2022) Examining the effect of information and communication technology, innovations, and renewable energy consumption on CO_2 emission: evidence from BRICS countries. Environ Sci Pollut Res 29:47696–47712. https://doi.org/10.1007/S11356-022-19283-Y/TABLES/7

84. Khan MK, Teng JZ, Khan MI, Khan MO (2019) Impact of globalization, economic factors and energy consumption on CO_2 emissions in Pakistan. Sci Total Environ 688:424–436. https://doi.org/10.1016/J.SCITOTENV.2019.06.065

85. Khattak SI, Ahmad M, Khan ZU, Khan A (2020) Exploring the impact of innovation, renewable energy consumption, and income on CO_2 emissions: new evidence from the BRICS economies. Environ Sci Pollut Res 27:13866–13881. https://doi.org/10.1007/S11356-020-07876-4/TABLES/11

86. Kim S (2022) The effects of information and communication technology, economic growth, trade openness, and renewable energy on CO_2 emissions in OECD countries. Energies 15:2517. https://doi.org/10.3390/EN15072517

87. Kocak E, Baglitas HH (2022) The path to sustainable municipal solid waste management: do human development, energy efficiency, and income inequality matter? Sustain Dev. https://doi.org/10.1002/SD.2361

88. Koçak E, Çelik B (2022) The nexus between access to energy, poverty reduction and PM2.5 in Sub-Saharan Africa: new evidence from the generalized method of moments estimators. Sci Total Environ 827:154377. https://doi.org/10.1016/J.SCITOTENV.2022.154377

89. Koçak E, Ulucak R, Dedeoğlu M, Ulucak ZŞ (2019) Is there a trade-off between sustainable society targets in Sub-Saharan Africa? Sustain Cities Soc 51: https://doi.org/10.1016/j.scs.2019.101705

90. Lau LS, Choong CK, Ng CF et al (2019) Is nuclear energy clean? Revisit of environmental Kuznets curve hypothesis in OECD countries. Econ Model 77:12–20. https://doi.org/10.1016/J.ECONMOD.2018.09.015

91. Li N, Ulucak R, Danish (2022) Turning points for environmental sustainability: the potential role of income inequality, human capital, and globalization. Environ Sci Pollut Res 29:40878–40892. https://doi.org/10.1007/S11356-021-18223-6/FIGURES/5

92. Li X, Ullah S (2022) Caring for the environment: how CO_2 emissions respond to human capital in BRICS economies? Environ Sci Pollut Res 29:18036–18046. https://doi.org/10.1007/S11356-021-17025-0/TABLES/5

93. Liu M, Ren X, Cheng C, Wang Z (2020) The role of globalization in CO2 emissions: a semi-parametric panel data analysis for G7. Sci Total Environ 718:137379. https://doi.org/10.1016/J.SCITOTENV.2020.137379

94. Liu X, Zhang S, Bae J (2017) The impact of renewable energy and agriculture on carbon dioxide emissions: investigating the environmental Kuznets curve in four selected ASEAN countries. J Clean Prod 164:1239–1247. https://doi.org/10.1016/J.JCLEPRO.2017.07.086

95. Liu Y, Sadiq F, Ali W, Kumail T (2022) Does tourism development, energy consumption, trade openness and economic growth matters for ecological footprint: testing the environmental Kuznets curve and pollution haven hypothesis for Pakistan. Energy 245:123208. https://doi.org/10.1016/J.ENERGY.2022.123208

96. Lorente DB, Álvarez-Herranz A (2016) Economic growth and energy regulation in the environmental Kuznets curve. Environ Sci Pollut Res 23:16478–16494. https://doi.org/10.1007/S11356-016-6773-3/FIGURES/6

97. Lykov IN, Obolenskaya EV (2021) Geoecological and economic assessment of the municipal solid waste management system in the Kaluga region. In: IOP conference series: earth and environmental science. IOP Publishing, p 12013

98. Magazzino C, Falcone PM (2022) Assessing the relationship among waste generation, wealth, and GHG emissions in Switzerland: some policy proposals for the optimization of the municipal solid waste in a circular economy perspective. J Clean Prod 351:131555. https://doi.org/10.1016/J.JCLEPRO.2022.131555

99. Magazzino C, Mele M, Schneider N (2020) The relationship between municipal solid waste and greenhouse gas emissions: evidence from Switzerland. Waste Manag 113:508–520. https://doi.org/10.1016/J.WASMAN.2020.05.033

100. Mahmood H, Maalel N, Zarrad O (2019) Trade openness and CO_2 emissions: evidence from Tunisia. Sustainability 11:3295. https://doi.org/10.3390/SU11123295

101. Martínez-Zarzoso I, Maruotti A (2011) The impact of urbanization on CO_2 emissions: evidence from developing countries. Ecol Econ 70:1344–1353. https://doi.org/10.1016/J.ECOLECON.2011.02.009

102. Mensah IA, Sun M, Gao C et al (2019) Analysis on the nexus of economic growth, fossil fuel energy consumption, CO_2 emissions and oil price in Africa based on a PMG panel ARDL approach. J Clean Prod 228:161–174. https://doi.org/10.1016/J.JCLEPRO.2019.04.281

103. Minghua Z, Xiumin F, Rovetta A et al (2009) Municipal solid waste management in Pudong New Area, China. Waste Manag 29:1227–1233. https://doi.org/10.1016/J.WASMAN.2008.07.016

104. Murshed M, Haseeb M, Alam MS (2022) The environmental Kuznets curve hypothesis for carbon and ecological footprints in South Asia: the role of renewable energy. Geo J 87:2345–2372. https://doi.org/10.1007/S10708-020-10370-6

105. OECD (2020) Environment at a glance indicators-circular economy-waste and materials. In: OECD
106. OECD (2018) Business models for the circular economy-opportunities and challenges from a policy perspective. In: OECD policy highlights
107. Ohlan R (2015) The impact of population density, energy consumption, economic growth and trade openness on CO_2 emissions in India. Nat Hazards 79:1409–1428. https://doi.org/10.1007/S11069-015-1898-0/FIGURES/5
108. Olanipekun IO, Olasehinde-Williams GO, Alao RO (2019) Agriculture and environmental degradation in Africa: the role of income. Sci Total Environ 692:60–67. https://doi.org/10.1016/J.SCITOTENV.2019.07.129
109. Olayungbo DO, Quadri A (2019) Remittances, financial development and economic growth in sub-Saharan African countries: evidence from a PMG-ARDL approach. Financ Innov 5:1–25. https://doi.org/10.1186/S40854-019-0122-8/FIGURES/2
110. Onifade ST (2022) Retrospecting on resource abundance in leading oil-producing African countries: how valid is the environmental Kuznets curve (EKC) hypothesis in a sectoral composition framework? Environ Sci Pollut Res 1:1–14. https://doi.org/10.1007/S11356-022-19575-3/FIGURES/3
111. Onifade ST, Erdoğan S, Alagöz M, Bekun FV (2021) Renewables as a pathway to environmental sustainability targets in the era of trade liberalization: empirical evidence from Turkey and the Caspian countries. Environ Sci Pollut Res 28:41663–41674. https://doi.org/10.1007/S11356-021-13684-1/FIGURES/3
112. Orlov A, Klyuchnikova E, Korppoo A (2021) Economic and environmental benefits from municipal solid waste recycling in the murmansk region. Sustainability 13:10927. https://doi.org/10.3390/SU131910927
113. Ouédraogo M, Peng D, Chen X, Hashmi SH (2022) Testing the mineral resources-induced environmental Kuznets curve hypothesis in Africa. Nat Resour Res 1–25. https://doi.org/10.1007/S11053-022-10060-9
114. Ozturk I, Al-Mulali U, Saboori B (2016) Investigating the environmental Kuznets curve hypothesis: the role of tourism and ecological footprint. Environ Sci Pollut Res 23:1916–1928. https://doi.org/10.1007/S11356-015-5447-X/TABLES/5
115. Park Y, Meng F, Baloch MA (2018) The effect of ICT, financial development, growth, and trade openness on CO_2 emissions: an empirical analysis. Environ Sci Pollut Res 25:30708–30719. https://doi.org/10.1007/S11356-018-3108-6/TABLES/7
116. Pasha A, Ramzan M (2019) Asymmetric impact of economic value-added dynamics on market value of stocks in Pakistan stock exchange, a new evidence from panel co-integration, FMOLS and DOLS. Cogent Bus Manag 6: https://doi.org/10.1080/23311975.2019.1653544
117. Pata UK (2021) Linking renewable energy, globalization, agriculture, CO2 emissions and ecological footprint in BRIC countries: a sustainability perspective. Renew Energy 173:197–208. https://doi.org/10.1016/J.RENENE.2021.03.125
118. Pata UK, Caglar AE (2021) Investigating the EKC hypothesis with renewable energy consumption, human capital, globalization and trade openness for China: evidence from augmented ARDL approach with a structural break. Energy 216:119220. https://doi.org/10.1016/J.ENERGY.2020.119220
119. Pedroni P (2000) Fully modified OLS for heterogeneous cointegrated panels. Adv Econom 15:93–130. https://doi.org/10.1016/S0731-9053(00)15004-2/FULL/XML
120. Pegkas P (2015) The impact of FDI on economic growth in Eurozone countries. J Econ Asymmetries 12:124–132. https://doi.org/10.1016/J.JECA.2015.05.001
121. Pesaran MH (2021) General diagnostic tests for cross-sectional dependence in panels. Empir Econ 60:13–50. https://doi.org/10.1007/S00181-020-01875-7/TABLES/11
122. Pesaran MH, Pesaran MH, Shin Y, Smith RP (1999) Pooled mean group estimation of dynamic heterogeneous panels. J Am Stat Assoc 94:621–634. https://doi.org/10.1080/01621459.1999.10474156
123. Pesaran MH, Ullah A, Yamagata T (2008) A bias-adjusted LM test of error cross-section independence. Econom J 11:105–127. https://doi.org/10.1111/J.1368-423X.2007.00227.X

124. Pham NM, Huynh TLD, Nasir MA (2020) Environmental consequences of population, afflu-ence and technological progress for European countries: a Malthusian view. J Environ Manage 260:110143. https://doi.org/10.1016/J.JENVMAN.2020.110143

125. Philippidis G, Sartori M, Ferrari E, M'Barek R (2019) Waste not, want not: a bio-economic impact assessment of household food waste reductions in the EU. Resour Conserv Recycl 146:514–522. https://doi.org/10.1016/J.RESCONREC.2019.04.016

126. Qiao H, Zheng F, Jiang H, Dong K (2019) The greenhouse effect of the agriculture-economic growth-renewable energy nexus: Evidence from G20 countries. Sci Total Environ 671:722–731. https://doi.org/10.1016/J.SCITOTENV.2019.03.336

127. Rafique MZ, Li Y, Larik AR, Monaheng MP (2020) The effects of FDI, technological inno-vation, and financial development on CO2 emissions: evidence from the BRICS countries. Environ Sci Pollut Res 27:23899–23913. https://doi.org/10.1007/S11356-020-08715-2/TAB LES/11

128. Rahman MM (2017) Do population density, economic growth, energy use and exports adversely affect environmental quality in Asian populous countries? Renew Sustain Energy Rev 77:506–514. https://doi.org/10.1016/J.RSER.2017.04.041

129. Razzaq A, Sharif A, Najmi A et al (2021) Dynamic and causality interrelationships from municipal solid waste recycling to economic growth, carbon emissions and energy effi-ciency using a novel bootstrapping autoregressive distributed lag. Resour Conserv Recycl 166:105372. https://doi.org/10.1016/J.RESCONREC.2020.105372

130. Renzhi N, Baek YJ (2020) Can financial inclusion be an effective mitigation measure? evidence from panel data analysis of the environmental Kuznets curve. Financ Res Lett 37:. https://doi.org/10.1016/J.FRL.2020.101725

131. Ridzuan S (2019) Inequality and the environmental Kuznets curve. J Clean Prod 228:1472–1481. https://doi.org/10.1016/J.JCLEPRO.2019.04.284

132. Sadik-Zada ER, Loewenstein W (2020) Drivers of CO_2-emissions in fossil fuel abundant settings: (Pooled) mean group and nonparametric panel analyses. Energies 13:3956. https://doi.org/10.3390/EN13153956

133. Sadorsky P (2014) The effect of urbanization on CO_2 emissions in emerging economies. Energy Econ 41:147–153. https://doi.org/10.1016/J.ENECO.2013.11.007

134. Salahuddin M, Alam K, Ozturk I (2016) The effects of Internet usage and economic growth on CO_2 emissions in OECD countries: a panel investigation. Renew Sustain Energy Rev 62:1226–1235. https://doi.org/10.1016/J.RSER.2016.04.018

135. Saqib M, Benhmad F (2020) Does ecological footprint matter for the shape of the envi-ronmental Kuznets curve? Evidence from European countries. Environ Sci Pollut Res Int 28:13634–13648. https://doi.org/10.1007/S11356-020-11517-1

136. Sarkodie SA, Adams S (2018) Renewable energy, nuclear energy, and environmental pollution: accounting for political institutional quality in South Africa. Sci Total Environ 643:1590–1601. https://doi.org/10.1016/J.SCITOTENV.2018.06.320

137. Sarkodie SA, Ozturk I (2020) Investigating the environmental Kuznets curve hypothesis in Kenya: a multivariate analysis. Renew Sustain Energy Rev 117:109481. https://doi.org/10.1016/J.RSER.2019.109481

138. Shafiei S, Salim RA (2014) Non-renewable and renewable energy consumption and CO_2 emissions in OECD countries: a comparative analysis. Energy Policy 66:547–556. https://doi.org/10.1016/J.ENPOL.2013.10.064

139. Shah SAA, Shah SQA, Tahir M (2022) Determinants of CO_2 emissions: exploring the unex-plored in low-income countries. Environ Sci Pollut Res 29:48276–48284. https://doi.org/10.1007/S11356-022-19319-3/TABLES/7

140. Shahbaz M, Nasreen S, Ahmed K, Hammoudeh S (2017) Trade openness–carbon emissions nexus: the importance of turning points of trade openness for country panels. Energy Econ 61:221–232. https://doi.org/10.1016/J.ENECO.2016.11.008

141. Shahbaz M, Topcu BA, Sarıgül SS, Vo XV (2021) The effect of financial development on renewable energy demand: the case of developing countries. Renew Energy 178:1370–1380. https://doi.org/10.1016/J.RENENE.2021.06.121

142. Sharif A, Raza SA, Ozturk I, Afshan S (2019) The dynamic relationship of renewable and nonrenewable energy consumption with carbon emission: a global study with the application of heterogeneous panel estimations. Renew Energy 133:685–691. https://doi.org/10.1016/J. RENENE.2018.10.052
143. Sharma KD, Jain S (2020) Municipal solid waste generation, composition, and management: the global scenario. Soc Responsib J 16:917–948. https://doi.org/10.1108/SRJ-06-2019-0210/ FULL/PDF
144. Sinha A, Shahbaz M (2018) Estimation of environmental Kuznets curve for CO_2 emission: role of renewable energy generation in India. Renew Energy 119:703–711. https://doi.org/10. 1016/J.RENENE.2017.12.058
145. Stern DI (2004) The rise and fall of the environmental Kuznets curve. World Dev 32:1419–1439. https://doi.org/10.1016/J.WORLDDEV.2004.03.004
146. Su M, Wang Q, Li R, Wang L (2022) Per capita renewable energy consumption in 116 countries: the effects of urbanization, industrialization, GDP, aging, and trade openness. Energy 254:124289. https://doi.org/10.1016/J.ENERGY.2022.124289
147. Tachie AK, Xingle L, Dauda L et al (2020) The influence of trade openness on environmental pollution in EU-18 countries. Environ Sci Pollut Res 27:35535–35555. https://doi.org/10. 1007/S11356-020-09718-9/TABLES/8
148. The World Bank (2019) Solid Waste Management. In: World Bank Br. https://www.worldb ank.org/en/topic/urbandevelopment/brief/solid-waste-management. Accessed 28 July 2022
149. Tiwari AK, Kocoglu M et al (2022) Hydropower, human capital, urbanization and ecological footprints nexus in China and Brazil: evidence from quantile ARDL. Environ Sci Pollut Res 2022 1:1–18. https://doi.org/10.1007/S11356-022-20320-Z
150. Tominac P, Aguirre-Villegas H, Sanford J et al (2021) Evaluating landfill diversion strategies for municipal organic waste management using environmental and economic factors. ACS Sustain Chem Eng 9:489–498. https://doi.org/10.1021/ACSSUSCHEMENG.0C07784/ SUPPL_FILE/SC0C07784_SI_001.PDF
151. Twerefou DK, Adusah-Poku F, Bekoe W (2016) An empirical examination of the environmental Kuznets curve hypothesis for carbon dioxide emissions in Ghana: an ARDL approach. Environ Socio-econ Stud 4:1–12. https://doi.org/10.1515/ENVIRON-2016-0019
152. UN (2022) Transforming our world: the 2030 Agenda for sustainable development. In: United Nations Dep. Econ. Soc. Aff. sustainable development. https://sdgs.un.org/2030ag enda. Accessed 29 July 2022
153. UNISAN (2022) What is a landfill? Why are landfills bad for the environment? | Unisan UK. https://www.unisanuk.com/what-is-a-landfill-why-are-landfills-bad-for-the-env ironment/. Accessed 28 July 2022
154. Uzar U (2021) The relationship between institutional quality and ecological footprint: is there a connection? Nat Resour Forum 45:380–396. https://doi.org/10.1111/1477-8947.12235
155. Van Tran N (2020) The environmental effects of trade openness in developing countries: conflict or cooperation? Environ Sci Pollut Res 27:19783–19797. https://doi.org/10.1007/ S11356-020-08352-9/TABLES/4
156. Vaverková MD, Adamcová D, Zloch J, et al (2018) Impact of municipal solid waste landfill on environment – a case study. J Ecol Eng 19:55–68. https://doi.org/10.12911/22998993/89664
157. Verbič M, Satrovic E, Muslija A (2021) Environmental Kuznets curve in Southeastern Europe: the role of urbanization and energy consumption. Environ Sci Pollut Res 28:57807–57817. https://doi.org/10.1007/S11356-021-14732-6/TABLES/8
158. Vollebergh HRJ, Dijkgraaf E, Melenberg B (2005) Environmental Kuznets curves for CO_2: heterogeneity versus homogeneity. Tilburg University, Center for Economic Research
159. Waheed R, Chang D, Sarwar S, Chen W (2018) Forest, agriculture, renewable energy, and CO_2 emission. J Clean Prod 172:4231–4238. https://doi.org/10.1016/J.JCLEPRO.2017.10.287
160. Wang Z, Bui Q, Zhang B et al (2021) The nexus between renewable energy consumption and human development in BRICS countries: The moderating role of public debt. Renew Energy 165:381–390. https://doi.org/10.1016/J.RENENE.2020.10.144

161. Wenlong Z, Tien NH, Sibghatullah A et al (2022) Impact of energy efficiency, technology innovation, institutional quality, and trade openness on greenhouse gas emissions in ten Asian economies. Environ Sci Pollut Res Int. https://doi.org/10.1007/S11356-022-20079-3

162. Westerlund J (2008) Panel cointegration tests of the Fisher effect. J Appl Econom 23:193–233. https://doi.org/10.1002/JAE.967

163. Wilts H, Bahn-Walkowiak B, Hoogeveen Y (2018) Waste prevention in Europe : policies, status and trends in reuse in 2017. In: European environment agency publications office

164. Wolde-Rufael Y, Weldemeskel EM (2020) Environmental policy stringency, renewable energy consumption and CO_2 emissions: panel cointegration analysis for BRIICTS countries. 101080/1543507520201779073; 17:568–582. https://doi.org/10.1080/15435075.2020.1779073

165. Yao Y, Ivanovski K, Inekwe J, Smyth R (2020) Human capital and CO_2 emissions in the long run. Energy Econ 91:104907. https://doi.org/10.1016/J.ENECO.2020.104907

166. Yu Y, Deng Y, Chen F (2018) Impact of population aging and industrial structure on CO2 emissions and emissions trend prediction in China. Atmos Pollut Res 9:446–454. https://doi.org/10.1016/J.APR.2017.11.008

167. Yuping L, Ramzan M, Xincheng L et al (2021) Determinants of carbon emissions in Argentina: the roles of renewable energy consumption and globalization. Energy Rep 7:4747–4760. https://doi.org/10.1016/J.EGYR.2021.07.065

168. Zafar MW, Mirza FM, Zaidi SAH, Hou F (2019) The nexus of renewable and nonrenewable energy consumption, trade openness, and CO2 emissions in the framework of EKC: evidence from emerging economies. Environ Sci Pollut Res 26:15162–15173. https://doi.org/10.1007/S11356-019-04912-W/TABLES/7

169. Zafar MW, Shahbaz M, Sinha A et al (2020) How renewable energy consumption contribute to environmental quality? The role of education in OECD countries. J Clean Prod 268:122149. https://doi.org/10.1016/J.JCLEPRO.2020.122149

170. Zafar MW, Sinha A, Ahmed Z et al (2021) Effects of biomass energy consumption on environmental quality: the role of education and technology in Asia-Pacific Economic Cooperation countries. Renew Sustain Energy Rev 142. https://doi.org/10.1016/J.RSER.2021.110868

171. Zaidi SAH, Danish HF, Mirza FM (2018) The role of renewable and non-renewable energy consumption in CO_2 emissions: a disaggregate analysis of Pakistan. Environ Sci Pollut Res 25:31616–31629. https://doi.org/10.1007/S11356-018-3059-Y/TABLES/8

172. Zaman Q uz, Wang Z, Zaman S, Rasool SF (2021) Investigating the nexus between education expenditure, female employers, renewable energy consumption and CO_2 emission: Evidence from China. J Clean Prod 312:127824. https://doi.org/10.1016/J.JCLEPRO.2021.127824

173. Zhang S, Liu X, Bae J (2017) Does trade openness affect CO2 emissions: evidence from ten newly industrialized countries? Environ Sci Pollut Res 24:17616–17625. https://doi.org/10.1007/S11356-017-9392-8/TABLES/8

174. Zhao W, Liu Y, Huang L (2022) Estimating environmental Kuznets Curve in the presence of eco-innovation and solar energy: An analysis of G-7 economies. Renew Energy 189:304–314. https://doi.org/10.1016/J.RENENE.2022.02.120

175. Zhu H, Duan L, Guo Y, Yu K (2016) The effects of FDI, economic growth and energy consumption on carbon emissions in ASEAN-5: Evidence from panel quantile regression. Econ Model 58:237–248. https://doi.org/10.1016/J.ECONMOD.2016.05.003

Ecological Footprint Assessment of e-Waste Recycling

Shameem Ahmad, Mohd Akram, Dilawar Husain, Akbar Ahmad, Manish Sharma, Ravi Prakash, and Mahboob Ahmed

Abstract Rapid technological advancements and socio-economic developments have given rise to the ever-increasing demands of electronic devices. These devices produce a huge amount of E-waste at the end of their operational life. According to the United Nations Global E-waste Monitor 2020 report, the annual generation of e-waste has grown up by 21% in the last five years. Recycling of e-waste is a sustainable measure to protect the ecosystem from its hazardous effects, but the recycling itself has a significant ecological impact on the planet. Prompt growth of e-waste recycling has created an urgent need for quantitative environmental assessment of e-waste recycling. The study proposed a methodology to assess the Ecological Footprint (EF) of e-waste recycling. The total EF_{ER} of E-waste recycling is estimated as 0.023 gha/tonne. The carbon absorption land contributes the highest among all

S. Ahmad
Department of Electronics and Telecommunication Engineering, Maulana Mukhtar Ahmad Nadvi Technical Campus, Malegaon 423203, India

M. Akram
Department of Computer Engineering, Maulana Mukhtar Ahmad Nadvi Technical Campus, Malegaon 423203, India

D. Husain (✉) · M. Ahmed
Department of Mechanical Engineering, Maulana Mukhtar Ahmad Nadvi Technical Campus, Malegaon 423203, India
e-mail: dilawar4friend@gmail.com

A. Ahmad
Department of Mechatronics Engineering, Symbiosis Skills and Professional University Pune, Pune, Maharashtra 412101, India

M. Sharma
Department of Mechanical Engineering, Malla Reddy Engineering College, Hyderabad, Telangana 500100, India

R. Prakash
Department of Mechanical Engineering, Motilal Nehru National Institute of Technology, Allahabad 211004, India

© The Author(s), under exclusive license to Springer Nature Singapore Pte Ltd. 2023
S. S. Muthu (ed.), *Environmental Assessment of Recycled Waste*,
Environmental Footprints and Eco-design of Products and Processes,
https://doi.org/10.1007/978-981-19-8323-8_5

other bio-productive lands for the e-waste recycling. Production and use of electronic devices will increase in near future and the e-waste recycling impact will also be scaled accordingly.

Keywords Recycling · e-waste · Ecological footprint · Waste management · Environmental assessment · Sustainable recycling

1   Introduction

The term electronic waste (e-waste) is used for the electrical and electronic devices which either have reached at the end of their operational life or due to some catastrophic damage it is not in usable condition. So, all the discarded electrical products come under the category of e-waste. It includes electrical home-appliances, domestic and industrial air conditioners, LED and other display devices, computers, and all types of personal gadgets like: Laptops, cellphones, tablets, printers, memory cards, cameras and gaming consoles. Rapid technological advancements and socio-economic developments have given rise to the ever-increasing demands of electronic devices. As a result of which the amount of e-waste produced every year is rapidly increasing. The annual production of e-waste has been increased by 21% in the last five years [1]. The e-waste is seen as a potential risk for the environment because it contains many toxic and hazardous substances like cadmium, lead, mercury, chromium etc. There are existing techniques to recycle this e-waste to save the environment from its ill effects. But the process of recycling has many complicated steps which have significant ecological impact on the planet. Improperly disposed e-waste releases heavy metals and toxic chemicals into the environment, which causes air water and soil pollution [2, 3]. The first problem is that a big amount of the waste is not even recycled in a proper manner. Devices like modern LEDs, circuits in small toys and other small devices generally end up in a dumpster[4–6]. In most developed countries the e-waste is disposed by simply disposing it in the landfills or incineration [7]. According to the United Nations Global e-waste Monitor 2017 report only 21% of total e-waste is recycled properly [8]. Though this percentage is increasing every year but still the recycling process itself needs to get more efficient so that the ecological impact of recycling on the environment can be reduced. But first we need to know more about the production of e-waste.

1.1   Categories of Appliances Generating e-Waste

The categorization of electronic appliances is shown in Fig. 1.

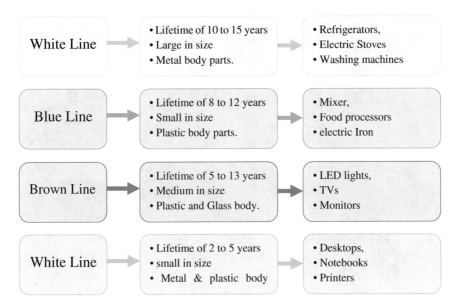

Fig. 1 Classification of electronic devices and appliances. Adapted from ABDI 2013

The white line has the longest operational lifetime i.e.,10 to15 years. It includes large appliances like electric stoves, refrigerators, dryers, air conditioners, dishwashers and washing machines. Most of the parts are metals and they have the least diversity of components.

The blue line of equipment includes small equipment which have an operational lifetime of 10 to 12 years, small equipment such as mixer, juicer coffee makers and blenders fall in this category. These appliances have plastic as its main compound. The brown line of equipment includes medium-sized equipment having an average life span of 5 to 13 years, such as televisions and tube, LCD, LED and PLASMA monitors, DVD, VHS players and camcorders. The main compounds of these appliances are glass and plastic.

The green line of equipment includes equipment with the shortest useful life (2–5 years). This is due to the programmed outmodedness and the large variety of components in its composition. So, the green line contains the type of equipment that impacts on the environment most. Mostly modern computers fall in this category like: notebooks, desktops, computers, tablets and cell phones.

1.2 Life Cycle Stages of e-Waste

Life cycle assessment (LCA) approach is used to assess and measure the impact on environment associated with different stages of creation, processing, and use of a product [9]. Life cycle assessment can be used to evaluate the environmental impact,

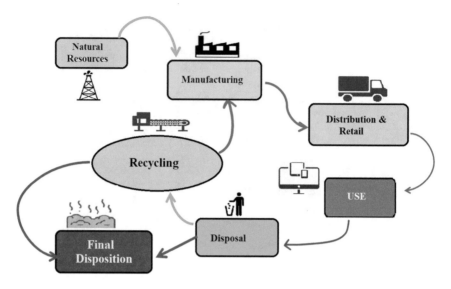

Fig. 2 Life cycle of electrical and electronic devices

inventory, key factors, optimization and decisions effectively and systematically and effectively. Numerous studies have evaluated the impact of e-waste treatment on the environment via LCA [10–12].

As can be seen from Fig. 2, there are different stages in the life of a device from natural resource to final disposition. The details are mentioned as follows:

Natural Resources

First the raw material from natural resources is collected to be delivered in the manufacturing units. These materials come from the mines, refineries and other manufacturing units. This includes plastic, metals and semiconductors.

Manufacture

The manufacturing units produce the final product to be sold in the market. This process itself generates a lot of environmental impact as the energy is being consumed and natural resources like water and papers are being exploited.

Distribution and Retail

Then the distribution to the retail and wholesale is done using various transportation vehicles. A large impact on the environment is due to distribution activities like transporting from one place to another.

User End

The user uses the equipment for their operational life time which varies according to the class of equipment.

Disposal

At the end of its operational life the equipment converts into e-waste and now there are two options: either to completely dispose it or to recycle it.

Recycling and Final Disposition

In recycling, the parts of equipment which can be used in other devices or can be dismantled to create raw materials that are extracted and the remaining parts which are completely un-usable are finally disposed. This is the stage where we want to focus. We have to understand the process of recycling further to accurately calculate the environmental impact of e-waste recycling.

1.3 Stages of e-Waste Recycling

E-waste recycling consists of three major stages:

i.   E-waste collection,
ii. mechanical pre-processing (device dismantling and/or shredding, sorting of e-waste fractions)
iii. end-processing (refining of e-waste fractions into raw material).

According to a report conducted by Cambridge Consultants, the key challenge faced by the recycling industry is that during mechanical preprocessing, the valuable target materials are also lost with junk into side streams. These valuable materials are unable to be recovered during end-processing also, which reduces the financial returns significantly [6].

For example, we consider Printed Circuit Boards (PCBs), where the PCBs contain 48% of the total monetary value as parts embedded in its fractions. Surprisingly, they account for only 8% of the total e-waste mass and this data stands valid across all types of e-waste [6].

The pre-processing can be conducted by two different ways:

(i) Destructive disassembly
(ii) Non-destructive disassembly.

1.3.1 Destructive Disassembly

In the destructive disassembly method, the whole device is broken down using a shredder and turned into fractions. After that, these shredded fractions are sorted so that materials can be separated from each other. Although, this method is convenient and provides high throughput, there are some massive drawbacks associated with it. The recovery rate of PCB fractions after shredding lies between 30 and 80%. Besides that, about 20% of the precious metals are lost due to mixing with non-recoverable materials such as plastics [6].

1.3.2 Non-destructive Disassembly

In the non-destructive disassembly method, first the device is disassembled and then each type of material is sorted and disposed separately. This type of disassembly method can achieve higher recovery rates. Since, the dismantling task is being performed before shredding, the valuable materials can be separated from nonrecoverable materials. If we use intensive manual disassembly to separate the parts, even 100% recovery can be achieved but, it is economically impractical because the overall throughput will be significantly decreased. Also, the smoke from these heavy metals can cause dangerous health problems to the human operators.

To maximize the efficiency of recycling and recovery of raw materials, shared optimization across this logistic chain is required. For example, we can reduce the End-of-Life devices into their material fractions: plastics, wiring, glass, printed circuit boards (PCBs), ferrous metals, etc. This is the key for the effective recovery of the embedded precious materials. Inefficiency of separation results in loss of material fractions into side streams from which they can never be recovered. For example, plastics fragments contaminating the purity of a collated PCB stream and vice versa.

1.4   Generation of e-Waste Globally

Generation of e-waste is rapidly growing continuously. It is because of the technological advancement and increasing online work and entertainment culture. According to the report published by United Nations, in 2019, the total quantity of globally generated e-waste was roughly 53.6 million metric ton (Mt) (excluding PV panels), which accounts to an avg of approximately 7.3 kg per capita. In 2010 it was only 33.8 million metric tons (Mts). Thus, we can say that the quantity of e-waste is growing at an alarming rate of almost 2 Mt per year globally [1, 13]. In 2019, the quantity of formally documented and recycled e-waste was 9.3 Mt, which is roughly 17.4% compared to the total e-waste generated. In 2014, formally documented and recycled e-waste was 1.8Mt. So, an annual growth of almost 0.4 Mt can be seen here. But this growth of 0.4Mt per year is very less as compared to the growth in the generation of e-waste which is almost 2Mt per year. This demonstrates that activities being done to recycle the e-waste are unable to keep up the pace with the global growth of e-waste generation [1].

The global quantity of e-waste is mainly comprised of large equipment (13.1 Mt), Small equipment (17.4 Mt) and Temperature exchange Equipment like ACs and refrigerators (10.8 Mt). Screens and monitors, Small IT and telecommunication equipment, and Lamps represent a smaller share of the e-waste generated 6.7 Mt, 4.7 Mt, and 0.9 Mt, respectively [1, 8]. Since 2014, the e-waste categories that have been increasing the most (in terms of total weight of e-waste generated) are the Temperature exchange equipment (with an annual average of 7%), Large equipment (+5%), and Lamps and Small equipment (+4%). This trend is driven by the growing consumption of these products in lower income countries, where the products enhance living standards. Small IT and telecommunication equipment have been growing at lower speed, and Screens and monitors have shown a slight decrease (−1%). This decline can be explained by the fact that, lately, heavy CRT monitors and screens have been replaced by lighter flat panel displays, resulting in a decrease of the total weight even as the number of pieces continue to grow.

Figure 3 shows the growth of global e-waste production and its recycled part from 2010 onwards [1, 13]. Generally, the recycling of e-waste is seen as the best option to dispose the waste and extract the essential raw materials to be used in manufacturing electrical equipment and other products, but the process of recycling itself has a significant impact on the environment. This study focusses on quantitative environmental assessment of global e-waste recycling. The study helps to understand the overall e-waste recycling process and provides the limit of recycling on the scale of sustainability. In this study, the Ecological Footprint analysis is used to measure the environmental impact of e-waste recycling on the planet. This study will also help to predict impact of e-waste recycling in near future and assist policy makers to take better decisions for the management and disposal of e-waste.

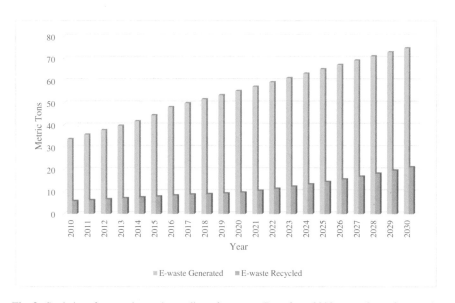

Fig. 3 Statistics of generation and recycling of e-waste. (Data from 2020 onwards are forecasts)

2 Methods and Materials

2.1 Ecological Footprint of e-Waste Recycling (EF_ER)

The case study illustrates the Ecological Footprint of E-waste recycling. E-waste recycling have significant environmental impact, therefore, to assess the environmental impact of E-waste recycling needs to calculate the impact of all processes that involve in recycling. The system boundary of the case study is shown in Fig. 4. The presented scenario of E-waste recycling contains energy consumption, water use, land use, labour requirement and transportation of E-waste materials to the treatment site and waste disposal to landfill etc. The EF assessment of E-waste recycling is as follows:

$$\text{EFER} = \sum \{E_i + \left(\frac{E_{mi}.D_{mi}}{E_{HDV}} + \frac{W_{wi}.D_{wi}}{E_{HDV}} \right) f_e + W_c.E_w.\} \alpha_i \frac{(1 - A_{oc})}{A_f} e_i$$
$$+ \sum (A_i - B_j) ei + \text{FTE.EF}_l$$

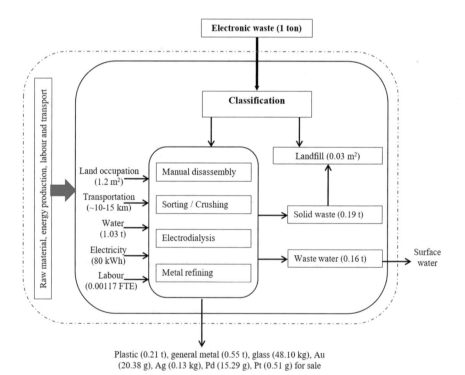

Plastic (0.21 t), general metal (0.55 t), glass (48.10 kg), Au (20.38 g), Ag (0.13 kg), Pd (15.29 g), Pt (0.51 g) for sale

Fig. 4 Resources involved in e-waste recycling for 1 metric ton

where,

E_i represents the amount of energy consumption for the treatment of E-waste (kWh), E_{mi} and D_{mi} are the amount of E-waste (tonne) and the average distance of transportation of E-waste materials to the treatment site (km), respectively. W_{wj} and D_{wj} are the amount of solid waste generation (tonne) and the average distance transportation of the solid waste to the landfill site (km), respectively. E_{HDV} represents the capacity of heavy-duty truck (HDV). α_i is the emission factor of the fuel (i.e., diesel, electricity etc.), f_e (kg of fuel per km transportation) is the average fuel efficiency of HDV. W_c represents water use in E-waste treatment (m^3) and E_w represents the energy required to uplift groundwater. α_i represents the emission factor of fuel/electricity (tCO_2/unit of fuel). A_{oc} is the fraction of annual oceanic emission sequestration, and A_f is the absorption factor of forests. e_i is equivalence factor (gha/ha) of different bio-productive lands. A_i represents the total occupied land (ha) for the E-waste treatment and B_j represents the landfill area (ha) required of the solid waste. FTE is full time equivalence of labour (1FTE = 8 h of working per day for one year or 250 days) requirement for E-waste recycling, EF_l represents the annual EF of labour/manpower. The details of different parameters are mentioned in Table 1.

The annual EF of labour assessed based on food consumption by the labour. The details of food goods consumption are mentioned in the Annexure Table 4.

Table 1 Parameters for the calculation of Ecological Footprint

S.No	Parameter	Value	References
1	E_{HDV}	3.5-tonne capacity	
2	f_e	0.240 kg/km	
3	E_w	0.14–0.23 kWh/m^3 of water uplift	[14]
4	α_i	Diesel: 3.17 kgCO$_2$/kg Electricity: 0.82 tCO$_2$/MWh	[15, 16]
5	A_{oc}	0.3	[17]
6	A_f	2.7 tCO$_2$/ha	[18]
7	e_i • Cropland ($e_{cropland}$) • Forest land ($e_{forestland}$) • Carbon absorption land ($e_{CO2\ land}$) • Marine/Sea-productive land ($e_{marine\ land}$) • Pasture land ($e_{pasture\ land}$)	2.52 gha/ha 1.28 gha/ha 1.28 gha/ha 0.35 gha/ha 0.43 gha/ha	

Table 2 The details of the different charges

Materials	*Cost (Rs)
E-waste	15,000 Rs/tonne
Labour cost	500 Rs/day
Transportation	18 Rs/tonne-km
Electricity	7.5 Rs/kWh

* All the data taken from local market

2.2 Economic Assessment

The economic assessment of E-waste recycling generally considers four parameters: (1) E-waste cost, (2) labour charges, (3) Machines operation charges, (4) Transportation cost. In this study all the charges are taken from the local market. The details of the different charges are mentioned in Table 2.

3 Results and Discussions

The results include the environmental impact of electricity use, transportation of E-waste, water use in recycling processes, land requirement and labour impact to assess the EF of E-waste recycling. The total EF_{ER} of E-waste recycling is estimated as 0.023 gha/tonne. The details of EF_{ER} assessment are as follows:

3.1 EF of Electricity Use

For one tonne of E-waste recycling, approximately 80 kWh electricity is consumed for crushing and shorting of E-waste. The Ecological Footprint of grid electricity is estimated by emissions factor (0.82 tCO_2/MWh) of electricity generation. The Ecological Footprint of electricity is 0.0003935 gha/kWh. The Ecological Footprint of electricity use for one tonne E-waste recycling is about 0.022 gha.

3.2 EF of Transportation

The transportation of E-waste recycling depends on two parameters:

(1) E-waste transport to recycling plant site and.
(2) transportation of solid waste to landfill site.

Some assumptions are adopted to estimate the transportational Ecological Footprint of E-waste. The Assumptions are as follows-

(a) *Considered 3.5-tonne capacity (E_{HDV}) HDV used for the transportation of E-waste & Solid waste.*

(b) *Distance of E-waste collection to recycling plant site (D_{mi}) should not more than 10–15 km.*

(c) *Distance of recycling plant to landfill site (D_{wj}) should not more than 10–15 km.*

The Ecological Footprint estimation considered emission factor of diesel fuel and fuel efficiency of HDV (capacity 3.5 tonne). The Ecological Footprint of transportation through HDV is estimated as 7.21×10^{-5} gha/tonne-km. The solid waste generation for one tonne of E-waste recycling is about 0.19 tonne. The Ecological Footprint of transportation is about 6.2×10^{-4} gha per tonne of E-waste recycling.

3.3   Water Use

Water use in E-waste recycling is significant because most of the countries are facing high-water stress. Groundwater is used most of the time in the recycling plant. Therefore, the electricity consumed for uplifting underground water is considered to estimate the impact of water use. The electricity consumption of water uplifting (up to 140 feet depth) is about 0.14–0.23 kWh/m^3 of water [14]. For one tonne E-waste recycling, nearly 1.03 tonne of water is required in the recycling plant. The Ecological Footprint of water use is about 5.2×10^{-5} gha.

3.4   Land

Generally, land requirement is not considered in environmental assessment of recycling. However, E-waste recycling required land to stock E-waste during recycling processes and also is needed for landfill of solid waste that was generated during recycling. Ecological Footprint analysis considered land occupation as a bio-productive land use; it depends on the type of land occupied (such as cropland, pasture land etc.) for any activity. The land occupancy for the one-tonne E-waste is about 1.2 m^2 (i.e., 1.2×10^{-3} ha) per tonne of E-waste recycling and 0.03 m^2 (i.e., 3×10^{-5} ha) per tonne of E-waste for landfill of solid waste. The Ecological Footprint of land occupancy for one tonne E-waste recycling is about 0.00031 gha.

3.5   Labour

In this study, the labour requirements for E-waste recycling are estimated in terms of FTE. The total labour required nearly 1.17 FTE for 1000 tonne of E-waste recycling

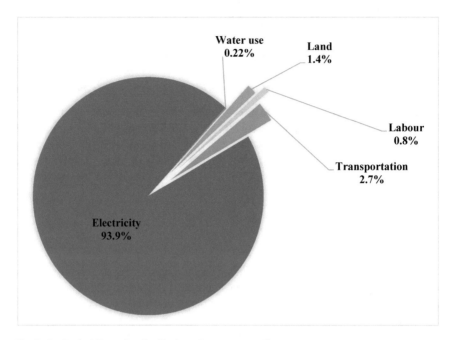

Fig. 5 Ecological Footprint distribution of e-waste recycling

[19]. The Ecological Footprint of labour is associated with the metabolic rate for different type of activities [20].

Figure 5 shows the Pei chart for different elements of Ecological Footprint of e-waste recycling. Here we can see that the highest impact is from electricity. Transportation is also contributing significantly as compared to land, labour and water use. The water used for recycling has the least impact on environment and contributes to only 0.22% of the total Ecological Footprint.

3.6 Economic Assessment

As discussed earlier, a large amount of metal resource is embedded in Electrical and electronics equipment that are being used. These metals could be recycled after the equipment has reached its End-of-life. Although the precious metals represent a small percentage of the e-waste (electronic scrap) weight, they are extracted from these devices due to their high monetary value. For example, in a mobile phone, the precious metals represent less than 0.5% of the weight but they contribute to over 80% of the monetary value, copper on the other hand, represents 5–15% of the economic value while having 10–20% share of the weight. The presence of precious metals also makes the recycling of less valuable elements feasible economically. For example, extraction of lead, tin, indium and ruthenium is economically feasible

Table 3 The costs of different quantities used in e-waste recycling

Items	Cost (Rs/Tonne)
E-waste	15,000
Electricity	600
Water use	1.43
Land	262.7
Labour	179
Transportation	214.2
Total cost of E-waste recycling	16,257.4 (~ 208.5 USD)

* At conversion rate of 1 USD = 78 Indian Rupees

because of the presence of other valuable elements such as gold, silver, palladium, and copper [21].

Numerous metals used in these equipment, specially rare earth metals (like: Ta, Co, Nb, In, Ga) are critical raw materials [22, 23]. These metals are not widely available throughout the globe. These metals are largely produced in very few countries, moreover some of them located in conflicted geographical areas which makes them more scarcer [24]. Generally, the concentrations of several metals are many times higher in e-waste than in natural metal ores which makes them desirable to be recycled [25]. Recycling these metals from e-waste helps to save resources such as energy and water [26]. In addition, for rare and precious metals the ecological footprint of recycling is much smaller than that of primary production [27].

It explains that recycling is essential because, on average, the reproduction of the same consumer electronic device requires around 10 times the final weight on raw resources [28]. Presently, however, most of the e-waste continues to be usually disposed off in sanitary landfills without proper recycling of these metals [29]. Market rate of E-waste, Electricity, Transportation and labour has been considered to calculate the cost of E-waste recycling. The total cost of one tonne E-waste recycling is calculated to be about Rs 16,257 (capacity 1000 Tonne Per Day of E-waste recycling). The local electricity rate is considered as Rs 7.5/kWh [30]. The costs of different quantities used in e-waste recycling are shown in Table 3.

At present, governments are becoming concerned about e-waste disposal because the increasing amount of e- more waste globally is creating a critical environmental problem for the planet. Even though the Life Cycle Assessment of e-waste has been studied extensively [31, 32], the potential impact of e-waste treatment on the environment varies widely [33].

3.7 Global Ecological Footprint of e-Waste Recycling

As the generation and recycling of e-waste keeps on increasing every year, the ecological footprint of its recycling process also grows annually. In this study we have

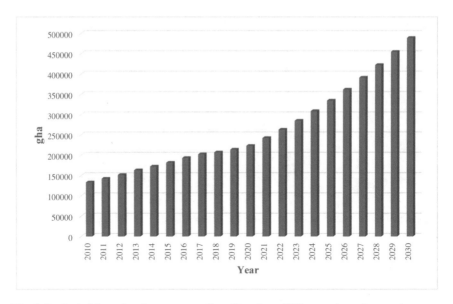

Fig. 6 Ecological footprint of e-waste recycling. (Data from 2020 onwards are forecasts)

calculated the EF of e-waste recycling for the last decade i.e., from 2010 to 2020. On the basis of this data, we have also predicted the EF of e-waste recycling for the next decade i.e., from 2021 to 2030. The graph in Fig. 6 shows that the EF is growing at an alarming rate. To reduce this growth, we have to optimize our recycling processes and create local recycling centres so that the EF due to transport can be reduced.

4 Conclusions

Recycling of E-waste is a sustainable measure to protect the ecosystem from its hazardous effects, but the recycling itself has a significant ecological impact on the planet. The study provides a novel method to assess the Ecological Footprint of E-waste recycling. Production and use of electronic devices will increase in the near future and the E-waste recycling impact will also be scaled accordingly. The Ecological Footprint of E-waste recycling is estimated as 0.023 global hectare per tonne (gha/t). The carbon absorption land contributes the highest among all other bio-productive lands for the E-waste recycling. An estimate of approximately Rs 16,257 (\$ 208.5) is to be spent to manage one tonne of e-waste in India.

It indicates that E-waste generation and E-waste recycling both damage the ecosystem of the planet, E-waste management also acts as a financial burden on the society. E-waste recycling can helpful for achieving United Nations Sustainable Development Goals. Hence, E-waste recycling percentage should increase as well as the demand for electronics devices should decrease in near future. The process of

Table 4 The details of ecological footprint of food per capita in India [34]

Food goods	Monthly consumption	Total annual consumption	CO_2 emission factor	Yield production (tonne/ha)	EF (gha/yr)
Vegetable	8.4 kg	100.8 kg	–	1.61	0.157
Pulses	0.90 kg	10.8 kg	–	0.69	0.039
Edible oil	0.85 kg	10.2 kg	–	0.38	0.068
Cereals	9.28 kg	111.36 kg	–	2.39	0.117
Fish	0.252 kg	3.024 kg	–	0.035	0.030
Mutton	0.08 kg	0.96 kg	–	72	0.006
Milk	5.4 L	64.8 L	–	458	0.062
Beef	0.06 kg	0.72 kg	–	32	0.011
Fruits	0.654 kg	7.848 kg	–	2330	1.58×10^{-6}
LPG	1.9 kg	22.8 kg	3.31($kgCO_2$/kg)		0.025
Wood	4.3 kg	51.6 kg	1.5–1.6 ($kgCO_2$/kg)	73 m^3/ha	0.028
Kerosene	0.40 L	4.8 L	2.58 ($kgCO_2$/litre)		0.004
Total annual EF/person					0.549

e-waste recycling must get more efficient to reduce the overall Ecological Footprint of E-waste recycling (Table 4).

References

1. Forti V, Baldé CP, Kuehr R, Bel G (2020) The global E-waste monitor 2020, no. July. United Nations University, 2020
2. Xu P et al (2014) Association of PCB, PBDE and PCDD/F body burdens with hormone levels for children in an e-waste dismantling area of Zhejiang Province, China. Sci Total Environ 499:55–61. https://doi.org/10.1016/j.scitotenv.2014.08.057
3. Li Y et al (Jun.2014) Polybrominated diphenyl ethers in e-waste: level and transfer in a typical e-waste recycling site in Shanghai, Eastern China. Waste Manag 34(6):1059–1065. https://doi.org/10.1016/j.wasman.2013.09.006
4. Raushan MA, Alam N, Ahmad S, Siddiqui MJ (2018) Improved performance of Dual material gate Junctionless TFET using asymmetric dual-k spacers, vol 1 pp 1–4
5. Ahmad S, Raushan MA, Kumar S, Dalela S, Siddiqui MJ, Alvi PA (2018) Modeling and simulation of GaN based QW LED for UV emission. Optik (Stuttg) 158:1334–1341. https://doi.org/10.1016/j.ijleo.2018.01.023
6. Biswas A, Husain D, Prakash R (2021) Life-cycle ecological footprint assessment of grid-connected rooftop solar PV system. Int J Sustain Eng 14(3):529–538. https://doi.org/10.1080/19397038.2020.1783719
7. Asante KA, Amoyaw-Osei Y, Agusa T (2022) E-waste recycling in Africa: risks and opportunities; Cu Husain D, Tewari K, Sharma M, Ahmad A, Prakash R (2022) Ecological footprint of multi-silicon photovoltaic module recycling. In: Muthu SS (ed) Environmental footprints

of recycled products. Environmental footprints and eco-design of products and processes. Springer, Singapore. https://doi.org/10.1007/978-981-16-8426-5_3

8. Balde CP, Forti V, Gray V, Kuehr R, Stegmann P (2017) The global e-waste monitor
9. "ISO - ISO 14040:2006 - Environmental management — Life cycle assessment — Principles and framework.". https://www.iso.org/standard/37456.html. Accessed 29 June 2022
10. Withanage SV, Habib K (2021) Life cycle assessment and material flow analysis: two under-utilized tools for informing e-waste management. Sustainability 13(14):7939. https://doi.org/10.3390/su13147939
11. Hong J, Shi W, Wang Y, Chen W, Li X (Apr.2015) Life cycle assessment of electronic waste treatment. Waste Manag 38(1):357–365. https://doi.org/10.1016/J.WASMAN.2014.12.022
12. Ismail H, Hanafiah MM (Jul.2021) Evaluation of e-waste management systems in Malaysia using life cycle assessment and material flow analysis. J Clean Prod 308:127358. https://doi.org/10.1016/J.JCLEPRO.2021.127358
13. Baldé C, Wang F, Kuehr R, Huisman J (2014) The global e-waste monitor
14. Plappally AK, Lienhard JHV (2012) Energy requirements for water production, treatment, end use, reclamation, and disposal. Renew Sustain Energy Rev 16(7):4818–4848. https://doi.org/10.1016/j.rser.2012.05.022
15. Eea, "EMEP/EEA air pollutant emission inventory guidebook 2013: Technical guidance to prepare national emission inventories," EEA Tech. Rep., no. 12/2013, p 23, 2013. http://www.eea.europa.eu/publications/emep-eea-guidebook-2013
16. (MPCEA), Ministry of Power Central Electricity Authority, "Government of India. (2016) CO_2 baseline database for the Indian power sector, user guide, 2016"
17. Monroe R (2013) How much CO_2 can the oceans take up? Scripps institution of oceanography. https://keelingcurve.ucsd.edu/2013/07/03/how-much-co2-can-the-oceans-take-up/. Accessed 30 June 2022
18. Husain D, Prakash R (2019) Life cycle ecological footprint assessment of an academic building. J Inst Eng Ser A 100(1):97–110. https://doi.org/10.1007/s40030-018-0334-3
19. United States Environmental Protection Agency (2020) Recycling Economic Information Report. https://www.epa.gov/smm/recycling-economic-information-rei-report#findings
20. Husain D, Prakash R (2019) Ecological footprint reduction of built envelope in India. J Build Eng 21:278–286. https://doi.org/10.1016/j.jobe.2018.10.018
21. Meskers C, Hagelüken C (2009) Green recycling of EEE: special and precious metal recovery from EEE. EPD Congr
22. Commission E et al (2018) Report on critical raw materials and the circular economy. Publications Office
23. E. Commission and J. R. Centre (2019) Recovery of critical and other raw materials from mining waste and landfills : state of play on existing practices. Publications Office
24. Wäger P, Wager PA (2011) Scarce metals - applications, supply risks and need for action. Not Polit XXVII(104):57–66
25. Szamałek K, Galos K (2016) Metals in Spent Mobile Phones (SMP) – a new challenge for mineral resources management. Gospod Surowcami Miner 32(4):45–58. https://doi.org/10.1515/gospo-2016-0037
26. Meskers CEM, Hagelüken C, Salhofer S, Spitzbart M (2009) Impact of pre-processing routes on precious metal recovery from PCs. Proc- Eur Metall Conf EMC 2:527–540
27. Hagelüken C (2008) Mining our computers -opportunities and challenges to recover scarce and valuable metals from end-of-life electronic devices. Electron Goes Green 2008 23
28. Kopacek B (2016) Intelligent disassembly of components from printed circuit boards to enable re-use and more efficient recovery of critical metals. IFAC-PapersOnLine 49(29):190–195. https://doi.org/10.1016/j.ifacol.2016.11.100
29. Bizzo W, Figueiredo R, de Andrade V (2014) Characterization of printed circuit boards for metal and energy recovery after milling and mechanical separation. Materials (Basel) 7(6):4555–4566. https://doi.org/10.3390/ma7064555
30. JAGMOHAN (2000) Manual on municipal solid waste management - 2000. http://cpheeo.gov.in/upload/uploadfiles/files/chap21(1).pdf

31. Song Q, Wang Z, Li J, Zeng X (2012) Life cycle assessment of TV sets in China: a case study of the impacts of CRT monitors. Waste Manag 32(10):1926–1936. https://doi.org/10.1016/j.wasman.2012.05.007
32. Niu R, Wang Z, Song Q, Li J (2012) LCA of scrap CRT display at various scenarios of treatment. Procedia Environ Sci 16:576–584. https://doi.org/10.1016/j.proenv.2012.10.079
33. Kiddee P, Naidu R, Wong MH (2013) Electronic waste management approaches: an overview. Waste Manag 33(5):1237–1250. https://doi.org/10.1016/j.wasman.2013.01.006
34. Dilawar Hu, Ravi P, Akbar A (2022) Life cycle ecological footprint reduction for a tropical building. Adv Civil Eng. Article ID 4181715, 14 p. https://doi.org/10.1155/2022/4181715

Ecological Footprint Assessment of Concrete Using e-Waste

Yakub Ansari, Dilawar Husain, Syed Mohammad Haadi, Umesh Kumar Das, Jyotirmoy Haloi, Khalid Iqbal, Ansari Abu Usama, and Ansari Ubaidurrahman

Abstract Rapid technological enhancement has become a cause of e-waste generation and will also create problems for the environment in near future. In this study, the environmental assessment of concrete using e-waste has been examined. The Ecological Footprint of the M20 grade plain cement concrete is calculated as 0.04678 gha/m^3, however, if 5–25% of aggregate (by volume) is replaced by e-waste in the concrete, the Ecological Footprint of the recycled plain cement concrete reduces by 0.12–0.59%. The partial replacement of the aggregate by e-waste in plain cement concrete reduces the concrete weight by 2.1–10.1%. The Sustainable Recycling Index (SRI) is also developed for the environmental assessment of e-waste as a replacement for aggregate in plain cement concrete. The SRI value of 5%, 10%, 15%, 20% and 25% aggregate replacement with e-waste are 1.6×10^{-5}, 6.4×10^{-4}, 1.4×10^{-4}, 2.6×10^{-4} and 4×10^{-4}; respectively. The production of concrete using e-waste provides a sustainable option for e-waste assimilation. Therefore, concrete using e-waste could be seen as an eco-friendly, lightweight, and economical alternative to the conventional concrete.

Keywords Ecological footprint · E-waste · Concrete · Waste management · Sustainable recycling index

Y. Ansari · U. K. Das · J. Haloi
Department of Civil Engineering, University of Engineering & Management, Jaipur 303807, India

D. Husain (✉)
Department of Mechanical Engineering, Maulana Mukhtar Ahmad Nadvi Technical Campus,
Mansoora, Malegaon, Nashik 423203, India
e-mail: dilawar4friend@gmail.com

S. M. Haadi · A. Ubaidurrahman
Department of Civil Engineering, SND College of Engineering and Research Centre,
Yeola 423401, India

K. Iqbal · A. A. Usama
Department of Civil Engineering, Maulana Mukhtar Ahmad Nadvi Technical Campus, Mansoora,
Malegaon, Nashik 423203, India

© The Author(s), under exclusive license to Springer Nature Singapore Pte Ltd. 2023
85

S. S. Muthu (ed.), *Environmental Assessment of Recycled Waste*,
Environmental Footprints and Eco-design of Products and Processes,
https://doi.org/10.1007/978-981-19-8323-8_6

1   Introduction

Urbanization and infrastructure development led to the demand for huge concrete production. The main reason behind its popularity is its high strength and durability [1]. Consumption of materials (like cement, fine and coarse aggregate etc.) in concrete is responsible for environmental problems when the extraction rate of sand, gravel, and other materials exceeds the generation rate of natural resources. Therefore, some alternative sources are used to replace materials in concrete [1, 2]. Electronic products have become an integral part of daily life which provides more comfort, security, and ease of exchange of information. The lifespan of different electronic products is mentioned in Table 1. These e-waste materials have serious human health concerns and require extreme care at their disposal to avoid any adverse impacts [2].

The electronics industry is the fastest growing and the largest manufacturing industry in the world [3]. Rapid technological advancement promotes the use of electronic products and simultaneously increases the rate of e-waste generation; e-waste generation rate is nearly three times faster than other solid waste in the world [4, 5]. The estimated global e-waste generation was 53.6 million metric tonnes in 2019 [6]. e-waste has become a problematic issue for humankind and the environmental options need to be considered especially recycling. However, e-waste recycling such as pyrometallurgy and hydrometallurgy are complex and expensive technologies to extract metallic and non-metallic components. The utilization of e-waste in plain cement concrete helps to recycle of them and also helps to reduce the manufacturing cost of concrete [7].

Some studies reported on e-waste use in plain cement concrete, Lakshmi and Nagan utilized e-waste as a replacement of coarse aggregate (up to 30%) in concrete and found a good compressive strength gain [9]. Suchithra et al. [9] stated that the addition of e-waste shows better compressive strength up to 15% replacement of aggregate in concrete. e-waste has more pronounced effect on the flexural strength than the split tensile strength [10]. Ahirwar et al. [10] stated that the workability of concrete increases with increase of e-waste percentage in concrete. The workability of fly ash with e-waste concrete gives better results than conventional concrete. However, the compressive strength decreased with increase of e-waste in concrete [11].

1.1   Plain Cement Concrete with e-Waste Use

Plain cement concrete consists of cement, fine aggregate and coarse aggregate. Details of the materials used in this study are mentioned as follows:

Cement Ordinary portland cement of 53 grades was used to prepare concrete samples. The cement has been tested for various proportions as per IS 4031–1988 and found to be conforming to various specifications of IS 12269–1987. The specific gravity was 2.96 and fineness was 3200 cm^2/ gm.

Table 1 Categories wise electrical and electronic equipment average lifespan [8]

Electrical and electronics equipment	Average Lifespan (year)
Information technology and telecommunication equipment	
Mainframe computer	10
Minicomputer	5
Personal computing: personal computers	6
Personal computing: laptop computers	5
Personal computing: notebook computers	5
Personal computing: notepad computers	5
Printers including cartridges	10
Copying equipment	8
Electrical and electronic typewriters	5
User terminals and systems	6
Facsimile	10
Telex	5
Telephones	9
Pay telephones	9
Cordless telephones	9
Cellular telephones	
Feature phones	7
Smart phones	5
Answering systems	5
Consumer electrical and electronics	
Television sets	9
Refrigerator	10
Washing machine	9
Air-conditioners	10
Fluorescent and other Mercury containing lamps	2

Coarse aggregate Crush angular metal of 20 mm size was used as coarse aggregate in concrete. The specific gravity and fineness modulus of used coarse aggregate were 2.71 and 7.31, respectively.

Fig. 1 e-waste (left side image) and e-waste after shredding (right side image).

Fine aggregate River sand was used as a fine aggregate in concrete. The specific gravity and fineness modulus of used fine aggregate were of 2.60 and 3.25, respectively.

e-waste electrical and electronic equipment such as computer, TV, Refrigerator, Air-Conditioning, radio etc. waste was used in plain cement concrete. However, e-waste was crush into small pieces (~10–20 mm size) in a grinder machine (see Fig. 1).

1.2 Procedure of Sample Preparation of Plain Cement Concrete

In the manufacturing of e-waste plain cement concrete (see Fig. 2), the following steps are followed:

1. Selection of materials and equipment is done before the starting of experimental work.
2. For testing, first the necessary thing is mixing of suitable proportions i.e., 1:1:2 of cement, fine and coarse aggregate with 0.4 to 0.45 proportion of water to prepare a wet concrete. As M20 grade plain concrete is used in this research work.
3. For checking the workability of the concrete slump value is tested and slump height is noted for all six types of mixture.
4. For casting of concrete cubes in mould with M20 grade, 5%, 10%, 15%, 20% and 25% of 10–20 mm e-waste chips, moulds are cleaned and checked for suitability in every aspect of the experiment. Then mixed concrete is poured into moulds.
5. After setting of concrete for one day in moulds, cubes are removed with precautions so that no damage take place.
6. All the specimens were prepared as per IS 4031–1988 standard.
7. All the cubes are dipped into a tank of water on the respective days of curing.
8. After the specified curing period every specimen is tested for compression Comparison graphs are plotted.

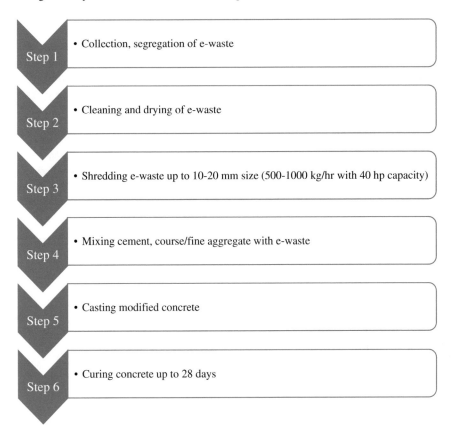

Fig. 2 Step-by-step of e-waste plain cement concrete manufacturing

9. The average value of the 3 specimens for each type is noted in the results and graph is plotted against it.

1.3   Limitation

The objective of the research work is to develop a method that can estimate the environmental impact of concrete. It does not estimate the lifecycle impact of concrete. The study also not consider the future degradation of bioproductive land during the calculations. The assumption in calculating the ecological footprint is the uniformity of bioproductivity of various types of lands, for example, forest land and cropland etc.

The study focuses on the Ecological Footprint assessment of plain cement concrete manufacturing and its reduction by using e-waste as a replacement of coarse aggregate. They made concrete specimens with partial replacement of e-waste on 5%,

10%, 15% and 20% to coarse aggregate. The compressive strength of the modified plain cement concretes has been experimentally examined and the results compared with the conventional M20 plain cement concrete. The study emphasizes to utilize e-waste as a building material and reduce the impact of the construction industry.

2 Methodology

The Ecological Footprint assessment gives an opportunity to assess the environmental impact of construction work on the planet. The method used in this study to compute the Ecological Footprint of e-waste assimilation is that of plain cement concrete. The details of Ecological Footprint assessment of e-waste in concrete are as follows:

2.1 Ecological Footprint of Concrete (EF_C)

In this study, three parameters are considered to assess the environmental impact of e-waste assimilation in plain cement concrete: (1) material use, (2) machinery/energy use and (3) labour involvement. The flow chart of the EF of e-waste assimilation in concrete manufacturing is depicted in Fig. 3. The Ecological Footprint of e-waste assimilation in concrete manufacturing has been estimated as Eq. (1):

$$EF_C = EF_m + EF_{ma} + EF_l \tag{1}$$

where,

EF_m represents the Ecological Footprint of materials use in concrete manufacturing; EF_{ma} represents the ecological footprint of machinery/energy use in concrete manufacturing. EF_l represents the Ecological Footprint of labour in concrete manufacturing.

2.1.1 EF of Material Use (EF_m)

EF_M is related to materials used during the concrete manufacturing. EF_m has been calculated by Eq. 2 [12]:

$$EE_m = \left(\frac{(Ci + Ri) \cdot Emi}{Af/(1 - Aoc)} \right) \cdot e_{CO2\,land} \tag{2}$$

where,

C_i and R_i are material consumption and waste generation of the ith material, respectively, E_{mi} is the embodied emission of the ith material. The average carbon absorption (A_f) of one hectare area of forests is about 2.7 tCO$_2$ [13], A_{oc} (i.e.,

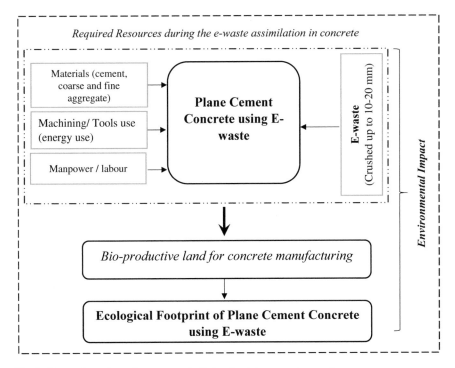

Fig. 3 System boundary of e-waste assimilation in plain cement concrete

Table 2 Equivalence factor of different bioproductive lands [15]

Bioproductive land	Equivalence factor e_i (gha/ha)
CO_2 absorption land ($e_{CO2\ land}$)	1.28
Forest land ($e_{forest\ land}$)	1.28
Crop land ($e_{cropland}$)	2.52
Pasture land ($e_{pasture\ land}$)	0.43
Sea productive/marine land ($e_{marine\ land}$)	0.35

0.30 [14]) is the fraction of annual oceanic emission sequestration. e_i represents the equivalence factor of different types of bio-productive lands as mentioned in Table 2.

2.1.2   EF of Energy/Machinery (EF$_{me}$)

EF$_{me}$ is related to machinery used in concrete manufacturing. EF$_{me}$ is calculated by using Eq. 3:

$$\text{EF}_{me}EF_e = \left(\sum C_e \cdot \alpha_e\right) \cdot \left(\frac{1 - A_{oc}}{A_f}\right) \cdot e_{CO_2\ land} \tag{3}$$

where,

C_e is the direct energy consumed in machinery; α_e is the emission factor of direct energy source.

2.1.3 EF of Labour (EF_l)

EF_L is considered the impact of labour involvement during concrete manufacturing. TEF_l directly depends on the metabolic rate of labour/manpower for a specific type of work [12, 15]. EF_l is determined by Eq. 4.

$$EF_l = FTE.EF_{me} \qquad (4)$$

where,

FTE is the full time equivalence, EF_{me} represents the Ecological Footprint of labour/manpower for unit FTE.

2.2 Economic Assessment

The cost of producing the M20 grade plain cement concrete and e-waste plain cement concrete is discussed in this study. Table 3 presents the cost details of different materials, machinery and labour requirement to produce different concretes. Market rate of the materials, machinery and labour has been taken to calculate the incurred cost of the produced concrete per cubic meter. Grid electricity cost was considered as 7.5 Rs/kWh for the estimation.

Table 3 Economic assessment of concrete manufacturing

Materials	Unit	Cost (Rs/unit)
Cement (PPC)	Kg	6.5
Water content	Kg	1
Sand/fine aggregate	Kg	0.7
Coarse aggregate	Kg	1.09
Water (Curing for 28 days)	Litre	1
Labour	Days	500
Concrete mixer	Days	800
Vibrator	Days	350
e-waste	Kg	5

2.3 Sustainable Recycling Index (SRI)

Various product-based sustainability indices are developed by various researchers for comparison of similar kind of products as well as they suggest which one is more sustainable in a certain boundary. To measure and compare the sustainability of e-waste assimilation in concretes, the Sustainability Recycling Index (SRI) has been developed based on two parameters: (1) Ecological Footprint of concrete and (2) cost of concrete. The SRI is a simple and effective tool to compare different types of mixed design concrete, it also suggests the limitations in mixed design concrete manufacturing. The expression of SRI is given in Eq. 5 [16]:

$$SRI = \left\{ \frac{EF_{con} - EF_{e-waste}}{EF_{con}} \right\} x \frac{C_{con} - C_{e-waste}}{C_{con}} \tag{5}$$

where,
   EF_{con} represents the Ecological Footprint of the conventional plain cement concrete; $EF_{e-waste}$ represents the Ecological Footprint of e-waste assimilation in the concrete; C_{con} represents the cost of the conventional plain cement concrete; $C_{e-waste}$ represents the cost of e-waste assimilation in the plain cement concrete.

Criteria for mixed design concrete manufacturing

The SRI value helps to develop criteria for concrete manufacturing. The three criteria for concrete manufacturing are as follows:

1. Sustainable concrete $0 < SRI \leq 1$;
2. Limit of recycling $SRI = 0$;
3. Unsustainable recycling $SRI < 0$

   The SRI value helps to understand the limit of e-waste mixing in concrete. For e-waste assimilation in concrete, the SI value lies in between 0 and 1, the e-waste assimilation in concrete would be sustainable in nature. However, for more than the unity value of SRI, the e-waste assimilation in concrete would become unsustainable. The unity SRI value indicates to limit of e-waste assimilation in concrete. For improving the sustainability of e-waste assimilation in concrete, the SRI value should be close to zero.

3 Results

The quantitative environmental assessment of e-waste concrete has been assessed by using Ecological Footprint analysis in this study. Comparative assessment was also done for the case building, if e-waste is used in the building's concrete work.

3.1 Ecological Footprint of Concrete Manufacturing

The Ecological Footprint of the M20 plain concrete is 0.04678 gha/m^3, however, for 5–25% of coarse aggregate replacement with e-waste, the Ecological Footprint of concrete reduced from 0.04669 to 0.04647 gha/m^3. The details of the e-waste concrete are mentioned in Fig. 4. The Ecological Footprint assessment of concrete with different amounts of e-waste are explained as follows.

3.1.1 EF of Material Use (EF$_m$)

The Ecological Footprint of concrete materials is estimated for different proposions of course aggregate replacement with e-waste in concrete. The Ecological Footprint of M20 Concrete materials is estimated to be 0.04361 gha/m^3 (i.e., 93.3% of the total Ecological Footprint of M20 Concrete). The e-waste addition in concrete reduces the materials impact, due to the low EF of the e-waste (0.0029 gha/t). The Ecological Footprint of concrete materials (5% coarse aggregate replacement with e-waste) is about 0.04356 gha/m^3 (i.e., 93.18% of the total Ecological Footprint of M20 Concrete). Similarly, the Ecological Footprint of materials for 10%, 15%, 20% and 25% coarse aggregate replacement with e-waste is 0.04351 gha/m^3, 0.04345 gha/m^3, 0.04339 gha/m^3 and 0.04334 gha/m^3, respectively. The details of Ecological Footprint distribution of concrete materials are depicted in Fig. 5 (Table 4).

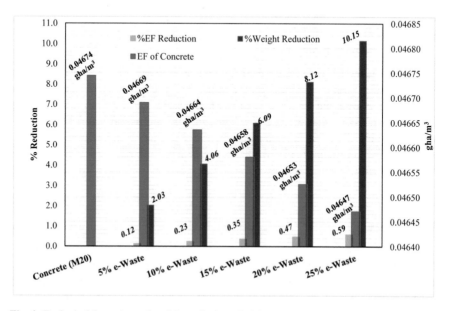

Fig. 4 Ecological footprint and weight reduction of plain cement concrete (M20) and modified concrete

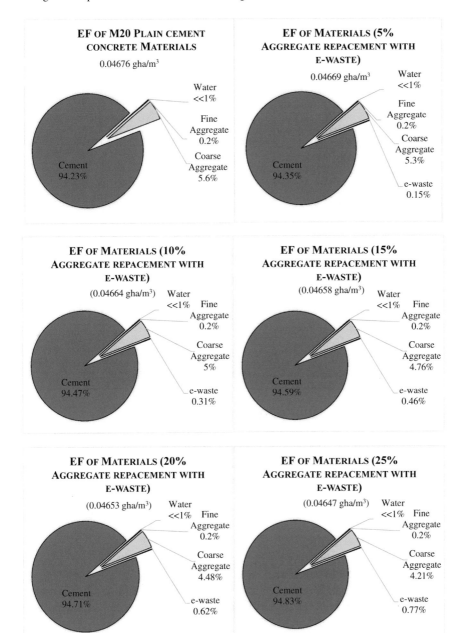

Fig. 5 Ecological Footprint distribution of concrete materials

Table 4 Ecological Footprint of materials and machinery

Materials	Unit	Quantity	Embodied energy (MJ/unit) [17]	tCO$_2$/unit [18]	EF (gha/unit)
Cement content	kg	397.96	5.32		1.032×10^{-4}
Water content	kg	197		0.01kWh/m^3 (up to depth of 36.5 m)	3.93×10^{-9}
Sand/fine aggregate	kg	663.27	464.29		0.012×10^{-5} [12]
Coarse aggregate -crushed stone	kg	1492.36	1626.67		0.00000163 [12]
Water for curing (for 28 days)	litre	168		0.01kWh/m3 (up to depth of 36.5 m)	3.93×10^{-9}
Labour	days	2.4 [19]			0.000901861
Machines mixer	days	0.07 [19]			0.00143068
Vibrator	days	0.07 [19]			0.0002932 [20]
e-waste (10 hp capacity machine required for shredding 500–1000 kg/hr)	tonne				0.00293

* Ecological Footprint of annual food consumption per capita 0.549 gha/yr [20, 21]

3.1.2   EF of Energy/Machinery (EF$_{me}$)

Machinery/Energy requirement for concrete manufacturing is estimated by the Central Public Works Department (CPWD) report (CPWD, 2016). The Ecological Footprint of machinery/energy requirement for concrete manufacturing (with or without e-waste) is same as 0.001 gha/m^3.

3.1.3   EF of Labour (EF$_l$)

The average labour-days (2.4 working-days for one m^3 concrete) requirment for the concrete is estimated by the Central Public Works Department (CPWD) report [19]. The Ecological Footprint of labour impact of concrete manufacturing (with or without e-waste) is almost same as 0.002 gha/m^3.

The distribution of the bioproductive land (material, energy & labour) for concrete (with or without e-waste) are shown in Table 5. Reduction of Ecological Footprint and weight of the plain cement concretes are listed in Table 6.

Table 5 Ecological footprint distribution of concrete

Concrete type	Materials (gha/m^3)	Energy (gha/m^3)	Labour (gha/m^3)	Total EF (gha/m^3)
M20 Plain Cement Concrete	0.04361	9.65×10^{-4}	2.16×10^{-3}	0.04674
Concrete (5% coarse aggregate replacement with e-waste)	0.04356	9.65×10^{-4}	2.16×10^{-3}	0.04669
Concrete (10% coarse aggregate replacement with e-waste)	0.04350	9.65×10^{-4}	2.16×10^{-3}	0.04664
Concrete (15% coarse aggregate replacement with e-waste)	0.04345	9.65×10^{-4}	2.16×10^{-3}	0.04658
Concrete (20% coarse aggregate replacement with e-waste)	0.04339	9.65×10^{-4}	2.16×10^{-3}	0.04653
Concrete (25% coarse aggregate replacement with e-waste)	0.04334	9.65×10^{-4}	2.16×10^{-3}	0.04647

Table 6 Ecological Footprint and weight reduction of modified concrete

Type of concrete	EF	% EF Reduction	% Weight Reduction
Plain Cement Concrete (M20)	0.04674	–	–
e-waste concrete with 5% coarse aggregate replacement	0.04669	0.12	2.03
e-waste concrete with 10% coarse aggregate replacement	0.04664	0.23	4.06
e-waste concrete with 15% coarse aggregate replacement	0.04658	0.35	6.09
e-waste concrete with 20% coarse aggregate replacement	0.04653	0.47	8.12
e-waste concrete with 25% coarse aggregate replacement	0.04647	0.59	10.15

3.2 *Economic Assessment of Concrete*

The manufacturing cost of the plain cement concrete (M20) is about Rs. 5923.5 per m^3, however the addition of e-waste decreases the concrete cost up to 6.9% of the conventional plain cement concrete cost. The cost of the modified e-waste concrete is mentioned in Fig. 6.

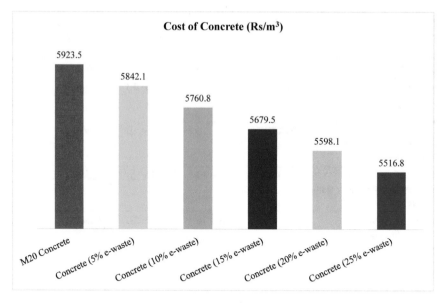

Fig. 6 Cost of the different modified concrete (Rs/m^3)

3.3 Sustainable Recycling Index (SRI)

The SRI value of the e-waste plain cement concrete with 5% e-waste used is estimated as 1.6×10^{-5}; it increases gradually with the increase of e-waste plastic percentage in the plain cement concrete. The SRI of all proposed e-waste concrete is depicted in Fig. 7. The SRI value of the e-waste concrete with 25% waste plastic is about 250 folds of the SRI value of the plain cement concrete with 5% e-waste.

3.4 Compressive Strength

After 28 days of curing, the strength of the concrete should be 100% that is 20.0 N/mm^2 of the M20 grade concrete. In this study it is found that replacement of the course aggregate with 15% e-waste by strength can be obtained until 19.8 which is approximately close to to 20 N/mm^2. Hence from this research work it can be concluded that 15% of e-waste can be replaced by the coarse aggregate in the plain cement concrete for the M20 grade. At the same time the density the concrete can be reduced which indirectly reduces the self-weight of the concrete member (Table 6). The compressive strength of the modified e-waste concrete is shown in Fig. 8.

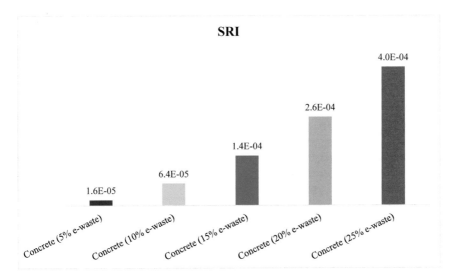

Fig. 7 Sustainable Recycling Index of different concretes

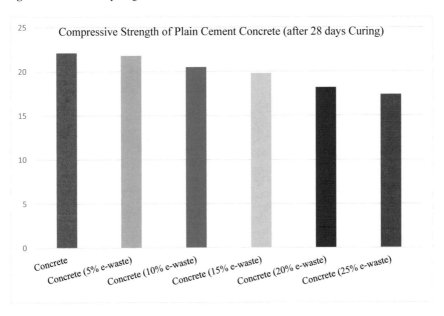

Fig. 8 Compressive strength of e-waste concrete

4 Conclusions

The Ecological Footprint of the plain cement concrete (M20 grade) and e-waste used plain cement concrete has been estimated and the results are compared with each other on the basis of sustainability scale. The Ecological Footprint of the conventional plain cement concrete (M20 grade) is about 0.04674 gha/m^3. The addition of e-waste in concrete decreases the Ecological Footprint of the concrete because a high environmental impact (coarse aggregate) material is replaced by a low environmental impact (e-waste) material. Replacing 25% of the coarse aggregate in the plain cement concrete with e-waste decreases the Ecological Footprint of the concrete up to 0.04647 gha/m^3 (i.e., 0.59% less environmental harmful than conventional plain cement concrete). However, aggregate replacement (i.e., 15% by weight) with e-waste gives a satisfactory compressive strength. Use of e-waste in the plain cement concrete also decreases the cost of the concrete. The SRI value of the concrete increases with increase of e-waste in the concrete.

The case study indicated that using e-waste in plain concrete is a sustainable option to reduce the demands of conventional building materials. e-waste can achieve the goal of sustainability if it is used efficiently in construction applications. Constructional Ecological Footprint can be reduced due to the low environmental impact of waste materials (i.e. e-waste, waste building materials etc.) and also helps to prevent global warming related to high concrete demand.

References

1. Alagusankareswari K, Kumar SS, Vignesh KB, Niyas KAH (2016) An experimental study on e-waste concrete. Indian J Sci Technol 9(2):1–15. https://dx.doi.org/https://doi.org/10.17485/ijst/2016/v9i2/86345
2. Shamili SR, Natarajan C, Karthikeyan J An overview of electronic waste as aggregate in concrete. https://doi.org/10.5281/zenodo.1132607
3. Clarke C, Williams ID, Turner DA (2019) Evaluating the carbon footprint of WEEE management in the UK. Resour Conserv Recycl 141:465–473
4. Nawandar V, Husain D, Prakash R (2021) Ecological footprint assessment and its reduction for packaging industry. Assess Ecol Footpr 41–78. https://doi.org/10.1080/19397038.2020.1783719
5. Husain D, Tewari K, Sharma M, Ahmad A, Prakash R (2022) Ecological footprint of multi-silicon photovoltaic module recycling. In: Muthu SS (ed) Environmental footprints of recycled products. Environmental footprints and eco-design of products and processes. Springer, Singapore. https://doi.org/10.1007/978-981-16-8426-5_3
6. Forti V, Baldé CP, Kuehr R, Bel G (2020) The global e-waste monitor 2020: quantities, flows, and the circular economy potential; United Nations University: Tokyo, Japan. ISBN 9789280891140
7. Kale SP, Pathan HI (2013) Recycling of demolished concrete and e-waste. Int J 260–268
8. Central Pollution Control Board (CPCB-2018). Ministry of environment, forest and climate change, government of India. https://cpcb.nic.in/uploads/Projects/E-Waste/e-waste_amendment_notification_06.04.2018.pdf
9. Lakshmi R, Nagan S (2010) Studies on concrete containing E-plastic waste. Int J Environ Sci 1(3)

10. Suchithra MK, Indu VS (2015) Study on replacement of coarse aggregate by e-waste in concrete. Int J Tech Res Appl 3(4)
11. Sunil A, Pratiksha M, Vikash P, Vikash Kumar Singh (2016) An experimental study on concrete by using E- waste as partial replacement for coarse aggregate. IJSTE - Int J Sci Technol Eng 3(4)
12. Husain D, Prakash R (2019) Ecological footprint reduction of built envelope in India. J Build Eng 21:278–286. https://doi.org/10.1016/j.jobe.2018.10.018
13. Mancini MS, Galli A, Niccolucci V, Lin D, Bastianoni S, Wackernagel M, Marchettini N (2016) Ecological footprint: refining the carbon footprint calculation. Ecol Indic 61(2):90–403
14. Global Footprint Network, (GFN) (2016) ⟨http://data.footprintnetwork.org/analyzeTrends.html?Cn=100&type=EFCtot⟩. Accessed November 2020
15. Scripps Institution of Oceanography (SIO) The Keeling Curve. https://scripps.ucsd.edu/programs/keelingcurve/2013/07/03/how-much-co2-can-the-oceans-take-up/. Accessed Aug 2020
16. Husain D, Garg P, Prakash R (2021) Ecological footprint assessment and its reduction for industrial food products. Int J Sustain Eng 14(1):26–38. https://doi.org/10.1080/19397038.2019.1665119
17. Abid AYZ, Usama AA, Husain D, Sharma M, Prakash R (2022) Ecological Footprint assessment of recycled asphalt pavement construction. In: Muthu SS (eds) Environmental footprints of recycled products. Environmental footprints and eco-design of products and processes. Springer, Singapore. https://doi.org/10.1007/978-981-16-8426-5_5
18. Inventory of Carbon & Energy (ICE). Sustainable Energy Research Team (SERT) (2011). www.bath.ac.uk/mech-eng/sert/embodied. Accessed Sept 2022
19. Plappally AK, Lienhard JH (2012) Energy requirements for water production, treatment, end use, reclamation, and disposal. Renew Sustain Energy Rev 16(7):4818–4848. ISSN 1364-0321. https://doi.org/10.1016/j.rser.2012.05.022
20. Central Public Works Department, Government of India, Analysis of Rates for Delhi (Volume-1&2) (2014). http://cpwd.-gov.in/Publication/DAR14-Vol1.pdf, http://cpwd.gov.in/Publication/DAR14-Vol2.pdf. Accessed 17 Aug 2021
21. Husain D, Prakash R (2019) Life cycle ecological footprint assessment of an academic building. J Inst Eng (India) Ser A 100(1):97–110. https://doi.org/10.1007/s40030-018-0334-3
22. Husain D, Prakash R, Ahmad A (2022) Life cycle ecological footprint reduction for a tropical building. Adv Civil Eng. Article ID 4181715, 14 p. https://doi.org/10.1155/2022/4181715

Printed in the United States
by Baker & Taylor Publisher Services